List of Illustrations

1) A photograph of Reginald Fessenden probably taken in 1905. 2
2) Helen Trott Fessenden in her early years. 5
3) A partial family gathering. 6
4). One of the laboratory work-rooms at Brant Rock. 22
5). The broadcast room at Brant Rock. 22
6). Artist's sketch of the 120 meter antenna at Brant Rock. 23
7). Fessenden and his staff at Brant Rock. 24
8). Schematic circuit diagram illustrating the U. S. Navy's use of the heterodyne principle in conjunction with an arc oscillator. 28
9). A 100 kilohertz converter from direct to alternating voltage of the type invented by Fessenden and marketed by the General Electric Company. 29
10) Schematic view of Fessenden's internal combustion engine cylinder. 40
11). A portion of a page from Fessenden's patent for means of generating powerful undersea acoustical waves. 43
12. Artist's conception of the operation of the Fathometer. 45
13. A mid-life sketch of Fessenden, probably from the period of World War I. 50
14). George Elliott, (Life Member IEEE), retired operations manager engaged in the design, manufacture and installation of gas turbine plants for electric power generation, gas and oil pumping, etc. across Canada. 57

Back cover: Tomb of Reginald and Helen Fessenden, Bermuda.

Contents

Introduction 1
       Background 3
       Education 8
Bermuda; New York; Thomas Edison 9
Newark and Pittsfield; Industrial Research 11
Academic Life (1892-1900): Purdue and Pittsburgh 14
Wireless Radio 15
The International Yacht Race; Marconi 16
The Weather Bureau (1900) 17
Wright Brothers 18
Trouble 18
Canadian Hopes 19
The National Electric Signaling Company 19
Continuation of Telephonic Transmission 26
The Turbo-Electric Drive 30
Legislation and Litigation 30
Navy Use of the Heterodyne System 34
The Superheterodyne System 36
Power Machinery and Power Storage 38
The Automobile Age 39
The Submarine Signal Company 41
World War I 45
More Frustrations 47
Geophysical Prospecting 49
Philosophical Issues 49
The "Research Council" 52
Aphorisms 54
Commentary 55
Inscription 61
Appendix I: Publications of Reginald Fessenden 62
Appendix II: Some Major Fessenden Patents 72
Index 75

To the memory of
Dr. Saul Dushman
Scientist, Friend, Councilor during my Schenectady days

INTRODUCTION

THE EXPLOITATION OF THE MAXWELL-HERTZ DISCOVERY of electromagnetic waves and the means of producing them with electrical equipment brought together a complex of individuals such as inventors, engineers, lawyers and entrepreneurs of various stripes, somewhat as in the case of a prospective gold mining venture. Guglielmo Marconi is the most celebrated of the group, partly because of the combination of temerity and speed with which he moved into the field, but several other individuals possessing great vision and talent, now almost forgotten, played at least comparable roles.

One of the greatest inventors of the period who merits much more recognition than he has commonly received is the generalist Reginald A. Fessenden (figure 1). He enjoyed close links to Canada, the United States and England. Although the core of his more public fame rests on his seminal contributions to wireless, the "firsts," contained in the papers he wrote and the more than 200 patents he was granted, cover an amazing range. These include, for example, the development of a chemical fire-retardant for electrical insulation and other materials, the microfilm slide, independent discovery of the special magnetic properties of the ferrosilicon alloys, promotion of the turbo-electric drive for ships, the very early design of x ray equipment for medical use, mechanized systems for parking automobiles, an electrically driven gyroscope, the undersea sound-producing equipment which led to sonar, geophysical prospecting with sound, and the location of hidden guns by sound ranging. He was also the inventor of a very light working air-cooled engine in which a sleeve attached to the piston is mounted outside rather than inside the cylinder in order to seal the combustion chamber. Some of the principles involved in it were probably employed later by others in the development of radial aircraft engines and in the engine situated in the rear of the first version of the Volkswagen automobile.

In addition to such important creations, Fessenden developed the concept of what is today termed amplitude modulated (AM) radio. He produced and improved upon equipment to demonstrate the principles involved, being the first individual to transmit voice and music over the air. He was the first to establish consistent two-way wireless communication across the Atlantic Ocean. In the course of the wireless phase of his work, he was

Figure 1) A photograph of Reginald Fessenden probably taken in 1905 during the development of the alternating potential generator for use in amplitude modulated transmission of sound. (Courtesy of the Fessenden collection of the Archives of the State of North Carolina in Raleigh, North Carolina).

granted a patent for use of the heterodyne principle that became so important in the vacuum tube era of radio and beyond.

Unlike Marconi, his choices of entrepreneurs could be unfortunate, requiring endless trying litigation. He sometimes took important steps forced upon him by his financial backers that were against his own better business judgment. However, his technical responses to such requirements were invariably brilliant since he had a highly creative mind and clear concepts of the ultimate goals to be achieved. While it has been said that some of his best concepts, such as the development of AM radio, were a decade or so ahead of their time, his successful demonstration of basic principles by "reduction to practice" provided major landmarks on the road to the future for others in the field.

Background

Fessenden was born in 1866 in East Bolton, Canada, on the shores of Lake Memphrémagog in Quebec, close to the Vermont border. His antecedents were colonists of English descent, the first member of the family, John Fessenden, having arrived in Cambridge, Massachusetts, in 1640.[1] The family was a distinguished one; one of the members, William Pitt Fessenden, served as Senator from Maine and as Secretary of the Treasury under Abraham Lincoln.

On the maternal side, Reginald's grandfather, Edward Trenholme, an English Protestant landowner, fled Ireland as a result of a life-threatening period of religious persecution in the 1820s. He settled in Canada where he owned and operated farms and gristmills. He died prematurely while in deep debt acquired while rebuilding one of the mills after a disastrous fire. This left his widow and descendants nearly penniless and struggling. One of his daughters, Clementina, married a young clerical student, Elisha Joseph

---

[1] Edwin Allan Fessenden has developed a two-volume history of the Fessenden family: *The Fessenden Family in America*, edited and completed by Mary Elizabeth Fessenden Washburn, privately published, 1971; Library of Congress Catalog Card No.,75-158409. This is available in the Genealogy Section of the main branch of the New York Public Library. Mrs. Washburn has speculated on the origin of the family name and concludes that it is most likely to be found in the Old Saxon-English language associated with the County of Kent in Southeastern England. Several interpretations have been suggested.

Fessenden, from a distinguished family. Reginald came into the world in the following year. In later days, Clementina became a leading activist for the creation of Empire Day within the British Commonwealth. She was very proud of her English heritage and was much afraid of the growing influence of life in United States upon Canadian culture.

Reginald had three younger brothers: Kenneth Harcourt, Cortez Ridley Trenholme and Victor Lionel, born respectively in 1868, 1870 and 1872.

There are a number of sources of information regarding the Reginald Fessendens. I have found several to be particularly useful. First, his devoted wife, Helen M. Fessenden (figures 2 and 3) who was a constant companion in all of his work during their years together, wrote an extended biography, *Fessenden; Builder of Tomorrows*,[2] after his death in 1932. Although she was not technically trained, she not only follows the main story closely with particular emphasis on general activities and many legal problems, but frequently provides valuable verbatim statements from technical documents written by others. I have leaned heavily her book as a guide to the sequence of events. It clearly reveals the deep affection the couple shared and the fact that she was a remarkable individual in her own right, being intimately involved in all the ups and downs of his professional career.

Actually, Fessenden did start to write a series of biographical articles for *Radio News*, one of the Hugo Gernsback publications that did so much to promote amateur interest in radio and television in the 1920s and 1930s. Unfortunately he was characteristically stiff-necked about accepting editorial suggestions (perhaps wisely) so that the series was terminated after a brief run and at about the time it would have become most interesting. Mrs. Fessenden incorporated the essence of this material into her biography.

I am especially indebted to Mr. George Elliott (figure 14) of Picton, Ontario, Canada, for correspondence and much other valuable material concerning Fessenden and his inventions. These include information from his own extensive professional engineering experience, from copies of the

---

[2]Helen M. Fessenden; *Builder of Tommorows*, Coward McCann Company, New York, 1940. I had the privilege of borrowing a copy held by the Brooklyn College Library. Mrs. Fessenden died in 1941, a year after the biography was published.

Figure 2) Helen Trott Fessenden in her early years. They were attracted to one another when they first met in Bermuda in 1884 and were married in 1890 when he obtained his first relatively stable position at a Westinghouse laboratory in Newark, New Jersey. She provided very strong moral and practical support to him throughout his career and prepared a detailed account of their complex adventures after his death in 1932. Courtesy of the Fessenden Trott Scholarship Committee, derived from the Fessenden Trott Trust in Bermuda. The photo was probably taken at the time of her graduation from college.

Figure 3) A partial family gathering. Clockwise starting from the upper left: Helen Trott Fessenden, Reginald Fessenden, Reginald Fessenden's mother, Clementina Trenholme, Reginald Kennelly Fessenden, probably age 3, the Reverend Elisha Joseph Fessenden, Reginald Fessenden's father. The photograph was probably taken in 1896 when Reginald and Helen lived in Pittsburgh. (Courtesy of the Fessenden collection of the Archives of the State of North Carolina in Raleigh, North Carolina).

bulletins of the now disbanded Amateur Radio Fessenden Society (ARFS), of which he was Editor, and from his informative essay "*Who was Fessenden.*"[3] In the latter Elliott casts light on the failings as well as the remarkable strengths of his subject. Several references to individual bulletins of the AFRS occur in the following text.

Ormand Raby, a Canadian, who was obviously fascinated with the range and brilliance of Fessenden's creations, wrote a popular book, *Radio's First Voice*,[4] which is somewhat complementary to Helen Fessenden's, although the author undoubtedly used her biography as a major guide to the sequence of events, much as I have done. In structure, his book actually takes the form of a popular novel that contains a continuous flow of imagined but plausible incidents involving conversation and action. Raby not only studied material from a number of archival sources and interviewed many individuals who had close connections with Fessenden personally, but visited sites in Canada and United States where he had been active.

Extracts from Raby's *Acknowledgments* are as follows:

> "Tracking down the facts of Fessenden's life required the better part of three years. Nearly one hundred persons (were) interviewed...
> "Numerous Fessendens and Trenholmes in the United States and Canada replied to my queries and to them, especially to Peter Fessenden, son of Cortez (Fessenden's brother), I owe a singular debt of gratitude.
> "The State Archives of North Carolina in Raleigh has the honor of caring for Reginald Fessenden's papers. Other important data are contained in the Clark Collection at the Smithsonian Institution in Washington and in the Douglas Library of Queen's University, Kingston, Ontario."

The Fessenden collection in the North Carolina State Archives referred to by Raby is particularly rich with respect to patents, published and unpublished papers, correspondence and photographs. Moreover the staff is highly cooperative.

---

[3] See: Proceedings of the Radio Club of America, Inc., Vol. 66, No. 3, p.25, 1992.

[4] Ormond Raby, *Radio's First Voice*, Macmillan of Canada, Toronto, 1970. In this case I had the privilege of a loan from the Boston University Library.

Beyond the foregoing is material available in the New York Public Library to which I had ready access. This includes extensive patent literature and many of the journals in which the some one hundred and fifty technically oriented papers that he wrote during his active years[5] were published. Some of the latter are listed in an appendix in Mrs. Fessenden's book and are reproduced at the end of this article, along with a list of some of his most important patents.

Since patents are usually prepared with the use of a skilled lawyer in order to impede infringement or to circumvent competing claims, they tend to have a special existence of their own and make a quasi-independent story. Fessenden's patents are generally brief, clearly written and informative. He was usually successful in defending his own patents, particularly those which dealt with his more uniquely advanced concepts. The professional papers he wrote during the peak of his inventive period deal, for the most part, with the results of his own research and development. Some describe recent advances; others are designed to protect priorities. In later years, much of his writing turned to philosophical and historical matters, as will be indicated later.

Education

At the age of nine he was admitted, on a fellowship and live-at-home basis, to the De Veaux Military College in Niagara Falls, Ontario, where his

---

[5]The branch of the New York Public Library on Madison Avenue has a complete record of the U. S. patent literature, including microfilm copies of the actual patents. It also has, in either bound or microfilm form, copies of the journals in which Fessenden published his technical papers. I am indebted to the library for the excellent service it provides. I am also most grateful to the staff of the Document Services of the Linda Hall Library of Kansas City, Missouri, for photocopies of *The U. S Patents of Reginld A. Fessenden* and *A New Reginald A. Fessenden Bibliography*, compiled respectively in 1990 and 1993 by David W. Kraeuter for the Pittsburgh Antique Radio Society, Inc. (Washington, Pennsylvania). The first bibliography lists his patents by subject, U. S. Patent Office number and year of issue. The second lists publications in addition to those contained in Helen Fessenden's book and other material. Similarly, generous help was provided by the staff in charge of the Fessenden Collection in the Archives of the State of North Carolina in Raleigh. Much valuable material, including photographs, correspondence, copies of published documents and patents is available there. Apparently the Fessenden's son Kennelly transferred the bulk of the family records to the Archive Center in Raleigh, North Carolina after his mother's death. Mrs Arwade spent three very profitable days inspecting the collection and gathering useful material. She is much indebted to J. R Lankford, Jr. and E. Morris for invaluable help.

father had a parish and his family resided at the time. He promptly became the star pupil. The precocious lad was much admired by his fellow students, though they were several years older than he and might instead have taken to hazing him. Two years later he left home for prestigious Trinity College School at Port Hope, Ontario. Again he won all honors while expanding his knowledge in the various directions the school offered, from languages and history to mathematics. Here he became familiar with the early versions of the *Scientific American*, then strongly oriented toward technical matters.

While waiting to enter college, he took a job at a bank but rapidly realized that a banker's life of that period did not suit his restless search for knowledge.

Finally, he completed his formal education at Bishop's College in Lennoxville, Quebec, majoring in mathematics. Here he found what was, for the time, a magnificent library. Among other things, he discovered the more scientific journal *Nature* which, with related scientific literature, opened a special world and led him into an entirely new line of interests.

Bermuda; New York; Thomas Edison

In 1884, at the age of eighteen, Fessenden left college before obtaining a final degree and was financially on his own. He was offered a position as headmaster at the Whitney Institute in Bermuda and accepted it. There he was well received by the local community. He opened up socially in an environment more relaxed than any he had previously known and met his future wife, Helen May Trott, to whom he then made a commitment. Bermuda was to become one of their anchor points in the complex life that followed. Although he enjoyed life on the islands enormously, he decided at the end of two years that he would move to New York City and seek employment in a scientific research laboratory, preferably that of Thomas Edison.

He bungled his initial application to Edison by understating his qualifications. To earn a living he had to be satisfied initially with the meager earnings he could obtain by writing articles on technology, economics, or social matters for popular journals. A break came, however, in that same year, 1886. The Edison Machine Works, which was laying electrical conduits in the streets of New York, offered him a position testing the quality of the interconnections. Incidentally, the first underground telephone lines in the city were being laid in the same trenches at that time. In the course of events, he met and retained a link with the prominent banker J. P. Morgan, who invited him into his palatial mansion to seek advice on the electric wiring in the residence.

His combination of high competence and dedication to work made him a marked individual. When the cable-laying program was completed, he was offered positions at either Schenectady or the Llewellyn Park Laboratory of Edison; without hesitation he accepted the second. In this manner there began a close, informal relationship between Fessenden and Edison that left a deep imprint in the mind of the young man. This was the period in which Edison, then in his early forties, was improving the electric lamp and phonograph and beginning development of the motion picture camera.

Fessenden was assigned to the chemistry division of the Edison laboratory and asked to see if he could find a form of electrical insulation for ordinary wiring that was flame-proof. Up to that point the insulation added to, rather than diminished, the hazard of fire. Following his own lead and testing in what might be called the true Edisonian manner (much more perspiration than inspiration), he found that an appropriate addition of antimony chloride to the insulating material provided a satisfactory solution. Edison was appropriately impressed and thereafter sought Fessenden's advice on many issues. Moreover, "Fezzy," as Edison called him, was raised to the rank of chief chemist as soon as the position could be made open. On one occasion Edison stated publicly in relation to Fessenden: "I can take a Yankee boy and a china mug and he will get more results than all the German chemists put together."

In Helen Fessenden's biography of her husband, she states that Fessenden's admiration for Edison as the leading inventor of the period remained unbounded throughout his life. However, he inevitably found it necessary to be guided more directly and to the extent possible by well-established theoretical principles, rather than trial and error. When he eventually moved into the field of wireless telegraphy and radio, his knowledge of mathematics and electromagnetic theory was indispensable.

What we today term theoretical chemistry was very primitive in the late 1880s. There were, in fact, very serious disputes among reputable scientists at that time as to whether atoms actually existed in the individual sense. There was no doubt in Fessenden's mind, however, and he conjured up his own views of what atoms and atomic forces must be like. Interestingly, he placed his theories of the nature of the atom and atomic forces among his highest contributions. Whatever else, his theories occasionally helped guide him to useful results.

In 1890, just before Edison left for a trip to Europe, Fessenden asked for and gained an agreement that he could work on wireless telegraphy when Edison returned. This was five years before young Marconi started his first work in Italy. Unfortunately, the financial backers decided that the company needed drastic refinancing and closed down the laboratory, leaving Fessenden footloose again at the age of twenty-four.

George Elliott (private communication) has expressed the opinion that the relatively abrupt termination of the Llewellyn Park Laboratory at a time when the young Fessenden was looking forward to following up on Hertz's experiments with much anticipation had a significant traumatic effect upon him. He became wary of highly organized entrepreneurial companies run by professional managers whose dominant, and perhaps only, interest in the laboratories they were financing lay, as he saw it, in the short-range possibilities for monetary return.

Newark and Pittsfield; Industrial Research

Among the openings offered to him in 1890 as he left the Edison Laboratory, he selected a position to work on the improvement of direct current motors and dynamos at the laboratories of the United States Company, a branch of Westinghouse located in Newark, New Jersey. He and Helen also consummated their long-delayed plans for marriage and settled into more normal domestic life. With a typical flair, Fessenden spent all of his savings at Tiffany's on wedding presents for Helen. Fortunately she had accumulated a small fund from her several years of teaching, permitting them to set up a modest household. Their only child, Reginald Kennelly, was born in 1893 (figure 3).

Their son's middle name was selected to honor Arthur Edwin Kennelly, a close friend since Fessenden's days with Edison. In 1902, Kennelly and

Oliver Heaviside first proposed, independently, the existence of an ionized layer containing free electrons in the upper atmosphere, the so-called ionosphere. It serves as a mirror for electromagnetic waves of sufficiently low frequency (below about thirty megahertz), making long-range transmission possible. Its existence was demonstrated by direct experiment in the 1920s.

Their son served in World War I and apparently suffered significant psychological trauma as a result. During a holiday visit to Bermuda at the end of World War II, he took to sea in a pleasure boat and did not return. No trace of him was found although the capsized boat was discovered.

The laboratory in Newark was somewhat run down at the time, but retained a memory of the fact that two great inventors, namely H. S. Maxim and J. A. Weston, had spent significant periods of their careers there. Maxim, along with other devices, had invented a machine gun; Weston had greatly improved standard electrical measuring equipment,

Among his many contributions to the improvement of direct current machinery was the discovery, based on his own atomic theories and applied research, that low-carbon ferrosilicon alloys have excellent magnetic properties. This includes good induced magnetization and relatively low energy loss in the cycling process involved in an alternating current transformer

He was also able to render a special service to the division of Westinghouse that was hoping to obtain an appreciable share of the market for electric light bulbs in competition with the Edison Company. At that time, the manufacturers were using relatively expensive platinum wires as conducting leads from the exterior of the bulb to the filaments in the lamp. Fessenden found that an inexpensive ferrosilicon alloy had the same thermal expansion coefficient as the glass commonly used for the bulb and served the same purpose well. This innovation, in addition to other improvements in manufacture that he proposed, permitted Westinghouse to gain the contract for lighting the Columbian Exposition that took place in Chicago in 1893. George Westinghouse was impressed.[6] In fact light bulbs had previously been

---

[6]The really major innovation in the production and operation of standard electric lamps at that period occurred when W. D. Coolidge of the General Electric Laboratory mastered the metallurgy of tungsten and the subsequent production of tungsten filaments, which were far superior to any predecessor. I once had the privilege of a private lunch with him in which he gave a detailed account of his research on tungsten, extending from metallurgical studies to the

sold at a price below cost in order to promote the use of electricity. Fessenden's contributions made this practice unnecessary.

I have found no evidence that Fessenden attempted to obtained a patent[7] on the use of silicon alloys for their magnetic properties. His first patent (No. 452,494, 1891) on their use as conduction leads for filament lamps indicates, however, that he was intimately familiar with the family of alloys as a result of his own work.

A Mr. Stanley, who leased research space in the same building as the laboratory and was part owner of a power plant in Pittsfield, Massachusetts, proposed that Fessenden join with him and two other experts on transformer design to create in Pittsfield what would be called the Stanley Company. Its goal would be to develop and market transformers. Fessenden agreed with the proposal. As a result he and Helen spent the winter of 1891-92 in Pittsfield. A national economic depression developed during that period, however, and the company failed.

The one reward he received from the venture was his first trip to England, made to study the status of electric power technology there. He was much impressed with Charles Parsons' development of the steam turbine, which he decided represented the wave of the future. He also had the pleasure of visiting Cambridge University where he met J. J. Thomson and J. A. Ewing, the ongoing expert on magnetism. Thomson showed him some of Maxwell's equipment.

In 1892, just as affairs at the Stanley Company looked darkest, he received an offer of a professorship in electrical engineering at Purdue University in Indiana which he promptly accepted.

---

actual production of filaments. He developed the practice of following his inventions from the most basic conceptual and bench investigations to manufacturing control. He later became the leader in the development of x-ray tubes.

[7] Credit for the discovery of the special magnetic properties of the ferrosilicon alloys is customarily given to Robert A Hadfield. See For example the book by F. Seitz and N. G. Einspruch, *Electronic Genie; The tangled History of Silicon*, p. 5, University of Illinois Press, 1998. However, Hadfield's memoir in *Obituary Notices of the Fellows of the Royal Society*, Vol. 3, p. 647, 1939-1947, contains the statement that the Hadfield group did not appreciate the special magnetic properties of the alloys until 1895, three or four year after Fessenden had evidently discovered and taken advantage of them on his own.

Academic Life (1892-1900);
Purdue and Pittsburgh.

Fessenden proved to be one of the most popular members of the Purdue faculty.[8] He was admired for his professional knowledge and standing, for his excellent lectures and his forthright open personality. After one year, however, he was lured to what is now the University of Pittsburgh by an offer much enhanced by the personal interest of George Westinghouse. This meant that he would have one foot in the dynamic industrial life of the city.

In addition to involvement with the university and local industry in Pittsburgh, Fessenden quickly formed close friendships with three notable individuals in the scientific world, namely John Brashear, James E. Keeler and Samuel Langley. The first had greatly advanced the art of grinding large lenses for astronomical and other purposes; the second was in charge of the local astronomical observatory; Langley, working under the sponsorship of the entrepreneur Harry Thaw, was among other things, studying the feasibility of developing a flying machine. The Fessendens formed a particularly close relationship with the Keelers and felt a substantial sense of loss when, in 1898, Keeler, for understandable reasons, accepted an appointment as director of the large and better situated Lick Observatory in California. One of his major discoveries there was the rotation of nebulae, a topic that has been of much interest to astrophysicists in recent years in the search for "missing" matter.

Roentgen's discovery of x-rays occurred during this period. When the Pittsburgh philanthropists, led by W. L. Scaife, learned of their potential use in medical diagnosis, they set up a fund to acquire a suitable x-ray unit for community service. Fessenden took on the responsibility for designing and constructing the equipment, being confident that it was the best available in

---

[8]An account of Fessenden's period at Purdue appears in the following paper: L. A. Geddes, *Fessenden and the Birth of Radiotelephony*, Proceedings of the Radio Club of America, Inc., Vol. 66, No. 3, p.20 1992. In this essay, Geddes mentions that the Department of Electrical Engineering of Purdue University received a bequest from the estate of Hellen Fessenden in 1980 to establish a fellowship in Fessenden's name. One concludes that the year at Purdue ranked among her most memorably happy ones. The Bank of Bermuda in Hamilton holds the trust funds which also provide fellowships to the University of Pittsburgh and a Canadian organization.

the country at the time. He succeeded in obtaining a patent on his design, although it was probably not commercialized.

He also became much interested in photography and its applications. Among other things, he developed equipment to make reduced copies of important letters, documents and drawings on individual segments of one-inch film which were then sandwiched and bound between comparably small flat glass plates. A matching, enlarging viewer was constructed as a companion device. Years later, in the 1920s, he noted that no one had commercialized, or even patented, a comparable system so he applied for and received patents on his latest version. It appears that he was at least the spiritual father of the very useful microfilm slide.

Most of the varied papers Fessenden published in professional journals during the academic years are what might be called standard fare, although some, such as one on flame retardation, went back to problems solved during his days with Edison.

Wireless Radio

As 1900 approached, however, matters relating to wireless telegraphy which was stirring interest in Europe, became prominent. He had begun to review the field in great depth. After much profound cogitation climaxed by inspiration, he became convinced that those working in the field had made a wrong turn and that he should take any reasonable opportunity to shift it in the right direction. In brief, he believed that the future lay in developing and exploiting continuous-wave transmission of electromagnetic waves having as nearly a single frequency as possible. Such waves could be modulated by an electromagnetic potential generated by a microphone and carry a continuous sound signal. Unfortunately the tools for generating such pure sinusoidal waves of arbitrary length were not readily available. He had to work with crude makeshifts, cobbled up from segments of damped Hertzian-type waves, but had just enough success to be convinced that he was on the right track.

It is to be noted that it was in this period that the Indian scientist J. C. Bose[9] commented in a scientific paper published in 1899 that he could hear

---

[9] J. C. Bose., Proceedings of the Royal Society of London, LXV, no. 416, p. 166, 27 April, 1899.

the passage of wireless pulses with good conventional earphones under proper circumstances. In response, or quite possibly in parallel, Fessenden proposed that the use of earphones be enhanced by rectifying the received signal. He anticipated that some components in the frequency band of the transformed wireless signal would lie in the most sensitive frequency band of the earphones. The rectifiers he used in the initial experiments were very inefficient, being either asymmetrical self-restoring coherers or point-contact diodes consisting of a metallic needle in contact with what was probably an inferior semiconductor. However, in 1902 he patented an efficient electrolytic rectifier consisting of a fine metal wire dipped in an electrolytic medium. Using his own special sense of appropriate nomenclature, he named it the "liquid barreter". It was widely adopted by wireless operators, but eventually replaced by more flexible crystal rectifiers.[10]

It is perhaps worth noting that many electrical engineers of this period who did not possess Fessenden's mathematical and theoretical background believed that the spark associated with the spark-gap switch introduced by Hertz was itself the source of the electromagnetic radiation, a misconception that greatly limited their ability to contribute to the development of the field. The point at which science began to lead the advance of technology had definitely been reached. It is hardly necessary to add that the noisy spark associated with wireless transmission dominated popular attention until vacuum tube electronics emerged into full view in the 1920s.

The International Yacht Race; Marconi

There was to be an international yacht race on the East Coast in 1899, generating much early excitement among sport fans. During the year previous to the race, Fessenden wrote to the editors of the *New York Herald* suggesting that they arrange to have a wireless-equipped craft move along with the racing boats and give a continuous account of the race as it progressed. The results would be received in the city and displayed continuously on public posters by the *Herald*. When asked if he were willing to take on the task, he referred the

---

[10] See, for example, the book mentioned in footnote 7. Also see: *The First Use of Crystal Rectifiers in Wireless,* Proceedings of the American Philosophical Society, p. 639, Vol. 142, 1998 by the same authors.

editors to Marconi who was very anxious to commercialize wireless at that early stage. Marconi, in turn, accepted the challenge. He tracked the race closely in a comparably fast boat, adding drama to the occasion. His transmission was a great success. It focused attention upon him as a glamorous individual and on the technology involved, as well as on the race. Moreover, it caused many to appreciate the potentialities of wireless for the first time.

In later years, Fessenden became highly critical of what he asserted were Marconi's unsubstantiated claims to priority in various aspects of wireless technology.

The Weather Bureau (1900)

In January of 1900, Fessenden received a letter from Willis L. Moore, Chief of the U. S. Weather Bureau, proposing that he join the staff of the Bureau and carry out wireless experiments of his own selection. The Bureau would supply money for equipment, supplies and special assistants, as well as members of its own staff. The initial research would be carried out at Cobb Island, sixty miles down the Potomac from Washington and available by steamer. Fessenden would be able to claim patents on inventions. This proposal was accepted and the move of the Fessenden family and staff took place in March. Thus they left the relative security of academic life behind.

Once settled into their somewhat primitive accommodations, they erected two antennae exactly a mile apart and plotted the detailed distribution of the transmitted radiation in the intervening area with considerable care. On one occasion a signal was sent to Arlington, Virginia, fifty miles away, to provide a modest demonstration of the capabilities of the system.

The most important experiment carried out on Cobb Island was a clear test, made in 1901, of the feasibility of voice transmission on the basis of amplitude modulation of a continuous wave. They had brought with them a 10,000 per second interrupter, produced in Pittsburgh to specification by Brashear, which made it possible to generate a more or less continuous wave, pieced together of one-tenth millisecond segments. This was modulated by voltage derived from a microphone; the signal was rectified at the receiving equipment. The received signal was quite noisy because of the irregular structure of the carrier, but there was no difficulty in distinguishing the transmitted words. Radio telephony was on the horizon! As mentioned earlier, a much more primitive experiment of the same kind had been tried successfully in Pittsburgh, presumably in 1899.

The individuals at the Weather Bureau were elated with the overall results. A decision was made to leave Cobb Island and establish wireless stations at Cape Henry, Virginia, Hatteras, and Roanoke Island, the latter two being in North Carolina. Headquarters were established at Roanoke Island. The equipment at Cobb Island was dismantled and sent there with some difficulty by ship. New equipment was procured for the other sites, which possessed operating weather stations. The investigations proceeded well through the summer of 1901 with communications between stations becoming routine. Moreover, there soon remained no doubt that telephonic communication was feasible between any two stations that could be linked by wireless waves.

The Wright Brothers

During this period, the Fessendens became close, lifetime friends of the Wright brothers, Orville and Wilbur, who were experimenting with gliders on the Carolina coast in preparation for their experiments on powered flight.

Trouble

By autumn, while contemplating possible business arrangements that might lead to the profitable sale of equipment, a somewhat jarring sign appeared on the horizon. The staff of the Weather Bureau showed evidence of wanting to take substantial control of the entire operation. Fortunately the matter remained relatively mute until well into the following year. In the meantime many demonstrations of the equipment were provided to individuals from other U.S. governmental agencies, representatives of foreign governments and presumably a few potential private customers.

The dormant problem with the Weather Bureau finally emerged in full force in the spring of 1902. The Chief of the Bureau insisted that he personally share the rights to patents which had been accumulated during the period that the work had been supported by the Bureau. In addition to being illegal in principle, the request was in complete violation of initial agreements.

Fessenden held his ground, protesting to Moore and his superior, then the Secretary of Agriculture, as well as to President Theodore Roosevelt. The Secretary, seeing the likelihood of a complicated wrangle, refused to take any action. In the meantime, Moore cut off all support derived from the Bureau, so that work came to a halt. Fessenden resigned his position as of September 1, 1902. Soon thereafter the Bureau closed down the three wireless stations and disposed of the equipment. Nevertheless, great strides had been taken.

Canadian Hopes

In searching for a new sponsor, Fessenden had hoped that he might form a strong link within his native land. In brief, he hoped that either the Canadian government or private entrepreneurs in the country would create a research and development center for wireless in which he would be the scientific and technical leader. The ultimate goal would be to establish a commercial wireless communication system and ultimately a sound-based radio network. A group of Canadian businessmen who appreciated the special stature Fessenden was attaining in the field was sufficiently interested in the proposition to join with him in forming what might be termed a shadow corporation which could expand into action when the time for commercialization was ripe. Unfortunately the ties between England and Canada eventually proved to be too strong; the Canadian government ultimately made an arrangement with the Marconi Company. For the present, Fessenden was thrown back on the dubious mercy of private entrepreneurs in United States.

The National Electric Signaling Company

Fessenden had stayed in constant touch with his Pittsburgh patent attorney, Darwin S. Wolcott, during all phases of his involvement with the Weather Bureau. In trouble, he called on Wolcott now for his best advice regarding the formation of a private company that would carry on additional research and development that might lead to the commercializing of equipment. In the process, he had discussions with individuals associated with a number of private organizations and had even been approached by one of his suppliers of electrical equipment, Queen and Company of Philadelphia, to form a joint organization. Wolcott, however, believed that a more adventuresome group of entrepreneurs could be found in Pittsburgh. Fessenden had established so many fine relationships with members of the Pittsburgh community during his eight years there that he was glad to follow Wolcott's advice.

Actually, an entrepreneurial organization in Pittsburgh based on Carnegie-Mellon financial interests came forward on its own and offered to support him on what presumably was a fair but conservative basis (see reference 4). In retrospect, he might have done well to accept this offer since he would then have been dealing with an experienced group linked to individuals having very broad vision and possessing large resources. As mentioned on page 11, George Elliott has proposed that Fessenden was probably reluctant to accept support from such a well organized professional group as a result of his painful experience associated with the abrupt closing of Edison's Llewellyn Park Laboratory by the managers of a similar organization 1890. In any event, he thought he could do better.

Wolcott ultimately turned up two sponsors, T. H. Given, an investment banker, and Hay Walker, Jr., a well established manufacturer, who were prepared to form with Fessenden and Wolcott what became the National Electric Signaling Company (NESCO). In the initial agreement, the sponsors would advance sufficient money to set up wireless stations for testing at Old Point Comfort and Cape Charles in Virginia and Bermuda. If the tests went well, the investors would have an option to buy 55 percent of the shares; the rest would be distributed between Fessenden and Wolcott as the former determined. Since it did not prove possible to establish a station in Bermuda, it and Cape Charles were replaced by Collingswood (near Philadephia) and Jersey City, both in New Jersey.

The tests, carried on in 1903, went well, but it soon became clear that Given and Walker would do their best to try to run the company as they saw fit. They respected Fessenden's technical genius and apparently hoped he would achieve results so dramatic that the company could be sold for a huge sum from which they would profit. The course that ensued proved to be a stormy one from start to finish. The entrepreneurs were primarily interested in expanding their wealth as rapidly as possible, whereas Fessenden, while interested in money, which would permit him to continue his work, was primarily focused on expanding the technological components of the base upon which civilization rests. Although the two goals were not entirely incompatible, the entrepreneurs sought for a quicker return than circumstances permitted; they had become involved in an enterprise that was beyond their capacity to handle properly. It should be emphasized, however, that the sponsors did raise about two million dollars, perhaps equivalent to eighty or so million today, to support the research program they preferred and that Fessenden responded magnificently when faced with a sensible technical challenge.

Very soon, Fessenden and his staff were prepared to demonstrate reliable transmitting and receiving systems that would operate over a distance of 150 or so miles. Unfortunately, circumstances did not make it possible to determine whether commercialization of such equipment would have been profitable. Instead the sponsors decided in 1905 that they should push for two-way transmission across the Atlantic Ocean, which had first been bridged in a relatively simple experimental test by Marconi in 1901. The sites ultimately chosen for stations were Brant Rock in Massachusetts, just north of Plymouth, and Machrihanish in western Scotland, with permission of the British government. The connecting great circle crossed the Bay of Fundy and Newfoundland, as well as a portion of the North Atlantic, for a distance of about 3,000 miles. (figures 4, 5,6 and 7).

Since relatively long wavelengths were to be used, tall antenna masts were needed. These consisted of a sequence of cylindrical steel tubes 400 feet high, placed on an insulated base and supported by four sets of four insulated guy wires. The method chosen to link the guy wires to the insulators was based on a reliable technique for connecting cables that had been developed by Roebling when constructing the Brooklyn Bridge. It was followed carefully by the contractor at Brant Rock. Unfortunately the same care was not taken by the staff of the same contractor in its work in Scotland. The antenna there was blown over within a year of being erected by a storm much less severe than some of those encountered at Brant Rock.

The two-way transatlantic system was ready for operation in the early part of 1906. In fact the first telegraphic signal from Brant Rock to Scotland

Figure 4). One of the laboratory work-rooms at Brant Rock. Fessenden is seated on the right. (Courtesy of the Clark Collection of the Smithsonian Institution and Dr. Frank. R. Millikan).

Figure 5). The broadcast room at Brant Rock. (Courtesy of the Clark Collection of the Smithsonian Institution and Dr. Frank R. Millikan).

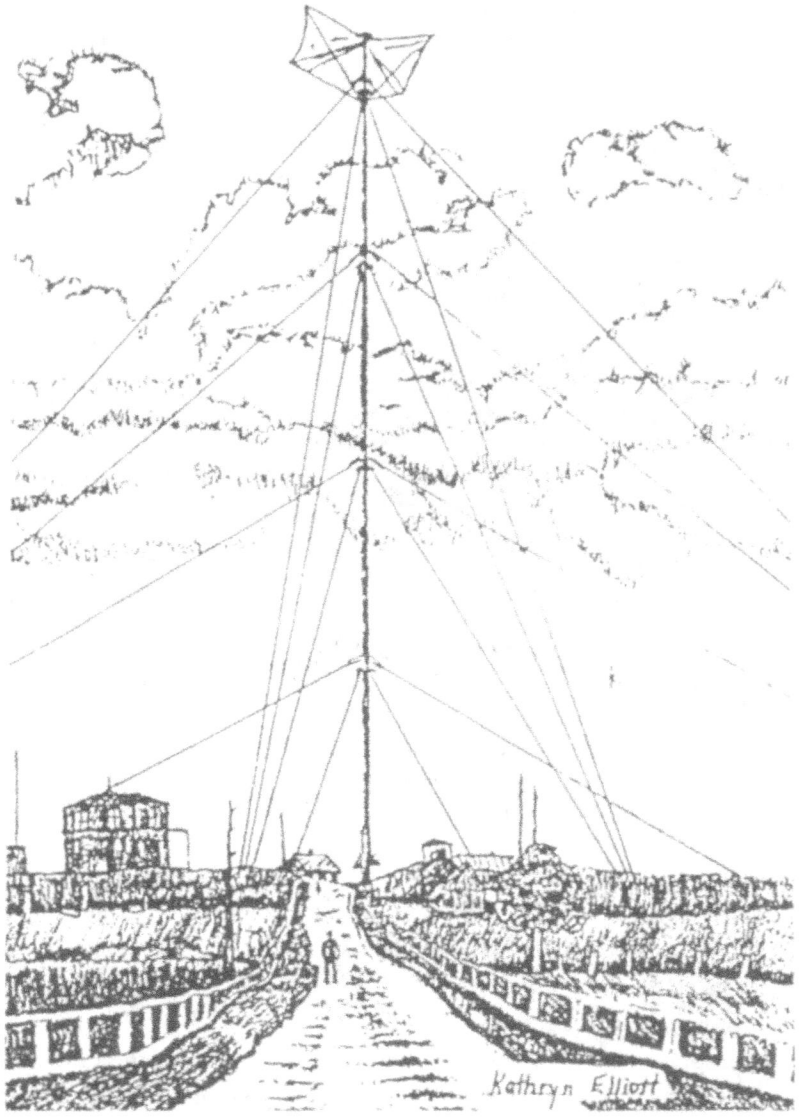

Figure 6). Artist's sketch of the 120 meter antenna at Brant Rock (right). It was duplicated in Scotland. (Courtesy of the artist, the daughter of George Elliott).

Figure 7). Fessenden and his staff at Brant Rock, presumably dressed for a special occasion. He is in front. The base of the antenna appears in the background. (Courtesy of the Clark Collection of the Smithsonian Institution and Dr. Frank R. Millikan).

was received on January 3rd; successful transmission in the reverse direction took place a week later. Although the system was never employed commercially, it served wonderfully for both experimental and demonstration purposes, being far in advance of any other at the time. It soon became clear that atmospheric conditions had a very large influence on the efficiency of transmission of the system at the wave lengths employed. The range in variation of the received signal from one period to another could be of the order of 1000, ultimately placing premium importance upon high transmission power supported by any technical "trick" that could enhance the audio portion of the received signal. The efficiency was particularly low during the daylight hours of summer.

In one especially fine period of transmission, the group at Brant Rock heard the "echo" signal that occurs about a fifth or so of a second after the first and is produced by a portion of the transmitted wave that has continued on and circumscribed the globe. On a similar occasion, an experimental test of voice transmission with use of the alternator and intended only for local reception was picked up in Scotland, much to everyone's surprise.

In the course of their work, the group found that transatlantic transmission was best for them in the range of frequencies below 70 kilohertz, corresponding to wavelengths above about 4 kilometers. Fessenden also noted, as a by-product, that transmission through air for distances much shorter than those involved in transatlantic reception was good in many regions of wavelength between 1 and 50 meters, later to be of interest for short wave communication and radar. Fessenden's measurements and decisions concerning the wavelength range to use for long distance transmission established something in the nature of an international standard until the early 1920s, when amateur operators carried out successful transatlantic broadcasts using equipment operating in the lower megahertz region.

Some of the nasty features of private competition in the field at the time are illustrated by an event that occurred during one such demonstration to an outside group. The entrepreneurs supporting the wireless work of Lee De Forest, who had invented the triode vacuum tube, hired an individual to operate a powerful arc in a laboratory building immediately adjacent to the Brant Rock test site, thereby making transmission and reception impossible. The individual in charge of the disturbing system was fortunately bought off with liquor and food. One is reminded of Mark Twain's comment about the goals of some entrepreneurs in this period of American business history: "Get rich, get rich quickly, get very rich, preferably by dishonest means, but by honest ones if no alternative is offered".

The fatal crash of the tower at Machrihanish occurred on December 5, 1906, after just less than a full year of two-way operation. A post-mortem study revealed the sources of weakness. Further research was to be centered at Brant Rock and confined to North America. The sponsors took the news in stride, presumably because they did not desire to commercialize transatlantic telegraphy. One cannot help but wonder why the crew working in Scotland, which included individuals from Brant Rock, did not inspect the links in the guy wires more carefully. The entire enterprise might have followed a different course had two-way transatlantic transmission continued.

Continuation of Telephonic Transmission

The primitive, but successful test of telephonic transmission at Cobb in 1901 encouraged Fessenden to pursue the issue with increased intensity.

In 1902 he had obtained major patents that dealt with what he termed the heterodyne principle, a concept essential for the future development of continuous wave technology as he envisioned it and which was indeed to have far-reaching and pervasive effects. If one passes two electromagnetic waves possessing different frequencies simultaneously through a non-linear element, such as a rectifier, one can generate two additional waves having the sum and difference frequencies of the initial pair. If one of the initial pair of waves has been modulated by the superposition of an electromagnetic signal derived, for example, from a sonic source, the two newly created electromagnetic waves will display the same modulation. Fessenden proposed "mixing" in this way a high frequency carrier bearing a sonic message with a wave whose frequency differs from that of the carrier by an amount lying in the sonic range. The undertone of the combination would then lie in the sonic range, permitting the sonic message to be heard in earphones or a loud speaker.

In his original primitive test of the principle, he transmitted two waves that differed by an audio frequency into earphones and was able to hear the undertone. In this instance his earphones and ears provided the necessary mixing.

Unfortunately he did not have the means to generate waves of arbitrary single frequency in 1902 so that the heterodyne principle was of limited, although not negligible, use to him in his immediate work. The patent did however become exceedingly important after the development of vacuum tube circuitry which made it possible to generate waves having a wide range of frequencies. Such sources at the receiving end, termed "local oscillators," could be used to alter the received signal, perhaps amplified, to a lower frequency, such as one in the audio range. The principle was also employed later in television and in radar. In the latter case, it was used to reduce the very high frequency of the received radar signal to a range that could be handled with conventional circuits. The patent faced many challenges over the years, but Fessenden's claims were always sustained.

He did much thinking about possible sources of relatively pure sinusoidal electromagnetic waves having frequencies well above the audible range that might be used as carriers of sonic messages, trying numerous options with variably satisfactory results.

Among other devices available to him was the oscillating arc, which was apparently invented independently by William Duddell and others, but which was improved extensively by the Danish inventor Vladimir Poulsen.[11] The latter, starting in 1902, developed and patented powerful versions which carry his name and which eventually proved to be exceedingly useful for the long range propagation of coded wireless messages (see page 34 et seq. for applications by the U. S. Navy). In addition to other components, the oscillating arc circuit contained a DC voltage generator and an arc which cycled in keeping with the complex resonant characteristics of the circuitry, including the behavior of the arc (figure 8). A significant part of Poulson's contributions to the development of high powered versions that operated at several hundred kilohertz lay in his discovery that the oscillations were sensitive to the nature of the gas surrounding the arc and the presence of a transverse magnetic field. Unfortunately the Poulsen arc produced a wider spectrum of frequencies than Fessenden desired.

Fessenden was well aware of the "singing arcs" as is clearly indicated in his written work. (See for example U. S. Patent 706,742.) He also used the oscillating arc as a local oscillator for heterodyne mixing in receivers designed for coded but not voice messages (See figure 8). Doubtless he followed the progress of Poulsen's research.

Finally, in 1906, Fessenden decided to focus on an electro-mechanical alternating current generator and placed an order with the General Electric

---

[11] See, for example, Hans Buhl, *Et Komparativt Studium af Poulsen-Systemets opfindelse, Udvikling og Inovation i Danmark, England og USA*. PhD dissertation, Natural Science Faculty, Aarhus University, Denmark, 1995. *(English Summary)*. I am indebted to Dr. Buhl for a copy of the booklet *The Telegraphone and the Arc Transmitter*, by Hans Buhl, Mark Clark, and Henry Nielsen, Danish Academy of Technical Sciences, English translation by Finn Jorgensen and Tina Jongla, 1998. Also see the book by J. H. Morecroft, A Pinto and W. A. Curry, *Principles of Radio Communication*, p. 580 et seq., John Wiley and Sons, New York, 1921.

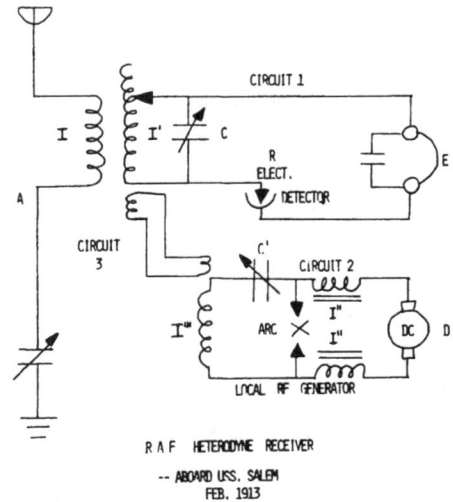

Figure 8). Schematic circuit diagram illustrating the U. S. Navy's use of the heterodyne principle in conjunction with an arc oscillator of the type developed by Poulsen in 1901 while searching for a source of continuous waves. It is based on earlier discoveries of "singing" arcs. This system served an exceedingly useful purpose for long range reception before and during World War I and until a wide variety of vacuum tubes became readily available. The vertical system A on the left is the antenna of a receiving system with an inductive link I to circuit 1 through a variable inductance I'. In the same circuit, C is a variable tuning condenser, E represents an earphone for reception and R is an electrolytic rectifier. Circuit 2 represents the arc oscillator. In the diagram, D is a direct current generator, C' is a variable condenser, the two inductances I" possess magnetic cores and X is the oscillating arc. The inductance I''' on the left side of circuit 2 can be coupled to the inductance I of the receiving antenna by way of the simple link represented by circuit 3, thereby "mixing" the frequency of the local oscillator with that of the received signal. It was found that the audible component of the received signal could be greatly increased by adjusting the frequency of the arc oscillator so as to be close to that of the main components of the received signal, thus producing a strong undertone signal in the audible range. Being much impressed with this success, the Navy not only adopted it for standard use but developed in parallel powerful versions of the arc generator for long range transmission of coded messages. Spark-switched oscillator generators of the type introduced by Hertz and used commercially by Marconi were abandoned as rapidly as possible. (Reproduced from Bulletin No. 33, October 1989, of the Amateur Radio Fessenden Society, Courtesy of George Elliott).

Company for one that would generate a 100 kilohertz wave. Although the best talent available in the company worked on the problem for two years, the device finally provided to him was limited to a range of 10 kilohertz. He decided to take on the problem with his own staff using in part the experience he had gained while working at the Westinghouse laboratory in Newark. He was eventually successful, devising a unit that produced about a kilowatt of radiation in the desired range of frequencies. Since he had great respect for the group in Schenectady, he transferred his own experience to its staff so that there would be an available source of such generators (figure 9). The individual who carried on the development there was E. F. W. Alexanderson, who eventually produced a 200 kilowatt version that could operate at 100 kilohertz. Years later, somewhat to his annoyance but perhaps even more to Helen's ire, the publicity department at the General Electric Company took complete credit for the entire development.

Figure 9). A 100 kilohertz converter from direct to alternating voltage of the type invented by Fessenden and marketed by the General Electric Company. The converter shown is one of Alexanderson's design. The driving direct current motor is on the right and the alternating current generator is on the left. The geared transmission is between the two. Alexanderson eventually produced a 200 kilowatt version that could operate at 100 kilohertz. (Courtesy of the Clark Collection of the Smithsonian Institution and Dr. Frank R. Millikan).

In any event, Fessenden was soon carrying on regular demonstrations of the new, greatly improved, system using Brant Rock and Plymouth as stations. High points of this period occurred on Christmas and New Year's Eve during the year-end holiday season of 1906. Turning up the power to full, he broadcast musical and vocal programs as broadly as possible. They were widely heard by wireless operators in ships along the coast since many had receivers using appropriate rectifiers. The Christmas program was picked up as far south as Norfolk, Virginia, that on New Year's Eve in the West Indies. By midsummer of 1907, telephonic communication was being carried on regularly between Brant Rock and Jamaica, Long Island, even during daylight hours. The stations were separated by about 200 miles.

The Turbo-Electric Drive

By the late 1880s, steam turbine systems were beginning to be employed for the propulsion of ships using mechanical transmissions. Fessenden had decided that it would be much more practical and efficient to use the shipboard turbine to generate electrical power and drive the propeller with an electric motor. He had proposed such a plan to the Navy in 1890. It in turn, recommended that he discuss the issue with the major electric companies. He presented the matter to both the Westinghouse and General Electric Companies, but they showed no interest on that occasion.

He pursued the concept relentlessly during the following years and finally, in 1909, succeeded in convincing both the Navy and the General Electric Company to develop the system. The company followed through and performed brilliantly. He sought compensation from it, pointing out his well-documented contribution to their success, but was rebuffed. Unfortunately his Boston lawyer died just before the case came to court and the suit was dropped.

Legislation and Litigation

Once it had become clear to governmental officials that wireless communications were here to stay and were of major importance, there was much discussion within their circles of enacting controlling legislation at both national and international levels. Action was not yet clearly focused, however, since many interests both public and private had to be consulted.

The uncertainties of the situation made it difficult for the Pittsburgh sponsors to find a possible buyer for the entire company as they would have wished. In 1907 Frederick P. Fish, the head of AT&T, seriously considered such a purchase since his board now believed after a long delay that the company's service should be extended nationally and not confined to the New York area. Just at that point, however, Fish was replaced by Theodore N. Vail who wished to study the issue as broadly as possible and develop relatively firm goals before moving forward. His capable technical staff decided to remain with wired systems. They purchased De Forest's patents on his triode vacuum tube and proceeded to improve the device during the following decade. Their success, matched by that of the General Electric Company under the leadership of Irving Langmuir, ushered in the age of vacuum tube electronics and with it an entirely new generation of creative scientists and engineers. The use of the heterodyne principle would soon come into its own and prove to have a pervasive influence on the course of major developments in electronics.

Since the main value of the National Electric Signaling Company lay in Fessenden's patents, the Pittsburghers attempted to gain control of them but failed after an extensive period of litigation. In the compromise agreement that emerged in 1908, Fessenden gained the right to sell equipment. With appropriate attention to marketing, in which he was highly capable, he might have been able to turn the company profitable and weather a complex period. For better or worse however he and the sponsors agreed that it would be wise for him to spend a period in London where legislative actions that could greatly restrict the international use of wireless were being reviewed intensively. In addition to becoming informed of trends and trying to influence their direction, he would study the opportunities for forming an English branch of the company.

The trip, made with Helen in 1910, managed unfortunately to stretch out for a full seven months. King Edward VII died soon after they arrived in England and they were swept into the center of the extensive formal funeral activities that took place. Moreover, since he was now one of the foremost leaders in wireless technology, his advice and opinions were sought from many sides; he gave as well as heard technical lecturers. He met with business competitors as well as possible allies and government officials. A constituency in Kent, the ancestral seat of the Fessendens, asked him to run for Parliament. And so it went. Above all, he felt that he had helped forestall the possibility that the British government would attempt to declare a monopoly on the use of wireless.

On returning to United States late in the year, he found the company in shambles. Those to whom he had given authority had handled it badly and were working at cross- purposes. Although he might have straightened out the situation in a month or two and started to do so immediately, the Pittsburghers were taking bold action. He was called to a meeting in Pittsburgh and then discharged from the company. The office at Brant Rock was to be terminated. The Company was placed in receivership. Samuel M. Kintner, a former student of Fessenden from Pittsburgh days who had meanwhile been employed by the Westinghouse Company, took on the problem of managing the residual affairs of the National Electric Signaling Company. He rapidly became Fessenden's antagonist, "sharper than a serpent's tooth". Actually, it is not clear who was really pulling the strings by this time since it had become a lawyer's game.

One basis used for discharging Fessenden lay in his continuing attempt to activate the wireless company he had helped create in Canada about the time the National Electric Signal Company was formed. Given and Walker asserted that the Canadian company should become a branch of their U. S. company. Fessenden refused, clearly feeling that his situation in the U. S. Company was becoming very insecure, in spite of his remarkable technical achievements; refuge in a productive independent Canadian organization could provide an excellent alternative.[12] As mentioned earlier, the Canadian government finally chose to work with the Marconi Company.

In order to be relieved of the time-consuming and psychologically painful burdens involved in the negotiations and to feel free to shift to new technical

---

[12] Fessenden's decision concerning the Canadian company is regarded to be somewhat reprehensible by some. See Bulletin No. 36, January 1990, of the Amateur Radio Fessenden Society, Box 737, Picton, Ontario K0K 2T0, Canada. As mentioned early in the text, the Society has been discontinued but I am indebted to Mr. George Elliott, its last editor, for much valuable information related to it.

problems, Fessenden did offer to turn all patents over to the company provided he would retain personally the right to use them. This occurred in 1916 during the war and at a time when the Marconi Company was very anxious to gain the rights to use some of the Fessenden patents. A nearly worthless verbal agreement not confirmed in writing was all that he achieved at that time. The plea of the Marconi Company for access to Fessenden's patents demonstrates again the great importance of his inventions for the advance of the field.

In brief, a very long period of litigation followed. Fessenden no longer controlled the rights to his most important patents and did not receive any remuneration from the company as agreed in the formal settlement of 1908. Matters were to drag on for nearly twenty years.

The company was sold to Westinghouse soon after World War I. Helen Fessenden states that the sum involved was "several million dollars". Westinghouse in turn transferred all its radio-associated patents, including those derived from Fessenden's work, to the Radio Corporation of America when it joined that organization which was formed in 1919 and based on a proposal by The General Electric Company that the major companies pool their patents. Westinghouse in turn received 3 million dollars for the Fessenden patents. At the time of the sale, the press in Pittsburgh lauded Given and Walker for their courage and farsightedness in having supported the field of wireless when it was still in the embryonic stage. While the commendations were in a sense deserved, the financiers might have fared better in the long run if it had been possible for them to gain a more intimate understanding of Fessenden and his motivations.

In 1922, Fessenden pressed claims against the new organization feeling much like David facing Goliath. Finally in 1928, perhaps partly to help mute some of the antitrust activity that had arisen in opposition to the patent pool and partly out of decency, RCA offered a reasonably generous settlement negotiated by Owen D. Young and his legal staff. This made it possible for the Fessendens to acquire a home in Bermuda and lead a more relaxing life for his few remaining years.

Had Fessenden been able to continue his research with radio another few years beyond 1911, he would have encountered the early stages of perfection of the vacuum tube and been faced with an entirely new series of opportunities and challenges as a result of the work of E. H. Armstrong and his counterparts in Europe on the development of vacuum tube oscillators and amplifiers. Considering the vigor with which he was able to pursue the field of underwater sonic research starting in 1912, it seems likely that Fessenden would have remained among the leaders in the new era of wireless for a period of time and possibly achieved more memorable and well deserved recognition in the field for what he had accomplished prior to the development of vacuum tube technology.

Before continuing with Fessenden's career and leaving wireless behind, however, it seems appropriate to present two areas of development in wireless technology which were affected in major way by his inventions in the field.[13]

Navy use of the Heterodyne System

In the years just prior to World War I, the U. S. Navy, which had established a radio research station at Arlington, Virginia, began experimenting with heterodyne systems. The naval engineers wished to see if it would be possible to enhance the distance over which coded wireless messages carried by continuous waves or trains of damped waves could be received by increasing the audible component of the message at the receiving end with the use of the heterodyne technology. The first tests employed the high frequency DC-AC converters developed by Fessenden and the General Electric Company. The marked success of the results obtained by conversion of the signal to the audible range led to rapid extension of the research.

As mentioned previously (page 27), Poulsen had devoted much attention to the further development of the somewhat cruder but practically useful form of continuous wave generator termed the variable frequency arc oscillator. Figure 8 illustrates a version of the oscillating arc circuit that Fessenden employed for heterodyne mixing when dealing with the reception of coded wireless. It contains two interconnected loop circuits. One branch of the

---

[13] In preparing the next two sections, I was greatly aided by the generosity of George Elliott for access to six selected Bulletins of the Amateur Radio Fessenden Society (see footnote 12), namely no. 29, April 1989; no. 33, October 1989; No. 34, November 1989; no. 35, December 1989; no. 36, January 1990; no. 37, February 1990. Elliott provides some interesting "What if?" comments concerning Fessenden's career in these Bulletins.

circuit consists of an inductance, a variable capacitor and a transformer that permits mixing with the received signal. The other portion of the circuit contains a direct current generator, two relatively large inductance coils and an arc interlinking and completing both loops. When the direct current generator is turned on, the arc begins to oscillate at a frequency determined by the adjustable electrical parameters of the system and the characteristics of the arc. Several modes of oscillation are possible, depending upon the conditions under which the arc is operated.

This device, enhanced by Poulson's important additions, was adopted for use by the U. S. Navy just before World War I and proved uniquely effective for long-range transoceanic telegraphy during the war. It could be employed as a local oscillator in a heterodyne mode at the receiver, as in the figure, or for the generation and transmission of a powerful continuous wave interrupted as needed for the transmission of code. The spark gap oscillator of the type initiated by Hertz would become obsolete. A number of shore-based stations that generated waves for very long range transoceanic communication employed the most powerful versions of the Poulsen arc available.

Actually, the vacuum tube oscillator, which was invented by E. H. Armstrong in 1912 and rapidly became the focus of much attention both nationally and internationally, soon demonstrated that it provides the best source of a mixing frequency and was used whenever available. Since manufacture of the new forms of equipment in substantial quantity took considerable time, some of the older, pre-vacuum tube, equipment saw extensive service throughout World War I. In fact, the Navy was compelled at times to manufacture electronic equipment for its own purposes because of the shortages of commercial sources.

Incidentally, the large text, *Principles of Radio Communication*, by J. H. Morecroft, A. Pinto, and W. A. Curry listed in reference 11 serves, perhaps unintentionally, as something in the nature of a valuable watershed document. It went to print in 1920, immediately after the war, marking the boundary between "old" wireless with its titanic struggles to establish basic principles and achieve incremental gains with marginally satisfactory equipment and the highly dynamic vacuum tube era that lay just ahead. It was written just before the advent of commercial radio broadcasting and the rise of a large generation of radio enthusiasts, ranging from amateur "hams" who assembled their own equipment to a widespread relatively passive but committed radio audience that depended on commercially manufactured receiving sets. The factors that would determine the economic base for the vacuum tube industry were still uncertain.

It is probably not unfair to say that the introduction of reliable vacuum tube technology had almost as revolutionary an influence on everyday life in its day as the introduction of the microchip is having in our own. It permitted broad and relatively pervasive use of the discoveries of Maxwell and Hertz.

The Superheterodyne System

The heterodyne principle came into wide use after World War I with the development of the superheterodyne system and the widespread emergence of popular commercial broadcasting. The master inventor in United States was E. H. Armstrong, who, as mentioned in the previous section, had developed means to control the oscillations of circuits containing triode vacuum tubes in the feed-back mode. The invention and basic work on the superheterodyne system began when Armstrong was involved in research in France during World War I and continued well into the 1920s. The characteristic features of the system can be illustrated by describing the operation of a relatively advanced AM radio receiver that employs heterodyne frequency converters at two places.

When a number of carrier waves of differing frequency, corresponding, for example, to waves transmitted from different commercial broadcasting stations, arrive at the receiving end, a receiving circuit may be tuned to resonate to a selected member of the group. A variable-frequency heterodyne converter, or *local mixing oscillator*, then converts the frequency of the selected signal to a predetermined *intermediate frequency* that is identical for all selected signals. The actual value of the intermediate frequency, chosen by the designer and manufacturer with considerations of the magnitude and quality of further amplification in mind, is well above the audio range. In North America, an intermediate frequency of 455 kilohertz was widely used by different manufacturers, but other values were employed both at home and abroad. In advanced versions of this system, the tuning condensers in the receiving and heterodyne circuits were coupled so that adjustment of only one tuning knob was needed to receive a signal from a given source and convert it to the specified intermediate frequency. The standard range of AM broadcasting frequencies in the 1920s lay in the band between 500 and 1500 kilohertz, not greatly different from the present one.

In a typical superheterodyne receiver, the received and converted signal next underwent several stages of amplification without change in frequency, the follow-on amplifier consisting of a chain of identical interlocked one-stage amplifiers. The number in the chain was variable; five stages were common.

Finally, the amplified intermediate frequency underwent conversion to the audio range with a second heterodyne mixer linked to an audio amplifier.

The superheterodyne system was subject to continuous evolution through the 1920s and mid-1930s, guided in part by the introduction of ever-more complex vacuum tubes. It ultimately found its place in the age of silicon chips following World War II. Its role in the development of radio communication immediately after World War I was so central that it became the focus of a virtual tornado of litigation. Fessenden played essentially no direct part in this personally since his interest lay, as mentioned previously, in procuring just settlements from RCA for his work with the National Electric Signaling Company. Samuel Kintner, who had returned to Westinghouse after the sale of NESCO was spokesman for the Fessenden patents.

In brief, the main controversies surrounding the superheterodyne system involved challenges to the validity of the Armstrong patents, then held by RCA, claims by RCA of infringement of those patents, and not unreasonable accusations that RCA's exclusive use of the patents amounted to a violation of the anti-trust laws. RCA in fact refused to license use of the superheterodyne system to any outside manufacturer during the period between 1924 and 1930. The excluded companies not only went to court, but also used all the technical ingenuity at their disposal to try to circumvent the monopoly. In 1930 RCA was compelled to issue licenses at a reasonable fee.

In the meantime Armstrong turned his attention to the development and marketing of a frequency modulated (FM) radio system that would operate at frequencies above those in the AM range; other individuals focused on television and radar.

Power Machinery and Power Storage

In 1905, early in his career with the National Electric Signaling Company, Fessenden was asked to serve as a technical member of what eventually became The Ontario Power Commission. At that time there was much interest on both sides of the border in deriving electric power from Niagara Falls. The immediate concern of the Commission was two-fold: To preserve the beauty of the Falls and develop power as effectively and efficiently as possible. He accepted the appointment with much pleasure. Moreover, activities associated with it eventually provided a more-than-welcome diversion from the worries arising from the signaling company as they worsened.

In addition to helping with the design of power-producing equipment, he proposed that water be conducted to the Falls by large pipes from the entrance to the Niagara River on Lake Erie, which was higher than the falls, thereby gaining the advantage of a greater pressure head. While the concept was quite sound, the available funds were insufficient to put his plan into service at that time.

There was growing interest in pumping water to reservoirs on the top of hills during periods of excess power production so that it could be used later to produce power during times of peak load. With characteristic imagination and inventiveness, he decided that there were circumstances in which it would be preferable to "invert" the process by storing the water in excavations or caverns one or two thousand feet below ground while extracting power, and pumping it back to surface reservoirs for later use during periods of power shortage. He received a patent on this concept in 1917, a decade after applying for one.

Together with the development of this concept, he explored efficient ways of using solar and wind power to assist in pumping water for storage and was again successful in obtaining patents. These novel ideas stimulated discussion at the time but were not put to practical use.

The Automobile Age

The automobile was proving to be very popular by 1911 and Fessenden, with time on his hands, decided to turn some of his attention to it. Parking of vehicles was creating a problem in older cities. He devised an automated parking system that would permit denser parking of cars in a given space than is possible if parked under their own power. Unfortunately the banks that had invested in parking areas were finding them unprofitable. He had arrived on the scene at the wrong time.

He also decided to develop an improved internal combustion engine and produced a novel design (figure 10) that weighed two pounds or less per horsepower, much lighter than any available airplane engine of the time. In it, the cylinder in which combustion occurred was a hollow tube with external grooves to carry lubricating fluid. The piston had two major components: first, a capped, hollow cylindrical core having a diameter somewhat less than that of the interior of the tubular combustion cylinder into which it fitted with sufficient clearance to avoid friction; second, an attached external cylindrical skirt or sleeve, coaxial with the core, having an inside diameter essentially the same as the outside diameter of the combustion cylinder, over which it could slide. Thus the second component of the piston was exterior to the combustion cylinder rather than inside. The tight fit of the external sleeve and the balance between its expansion and that of the combustion cylinder during operation eliminated need for the sealing rings employed on pistons in the conventional internal combustion engine. The engine could be used in either the two- or four-cycle mode and operate without spark plugs, that is in the diesel mode. In his patent, Fessenden envisioned an example, among others, in which two such engines operated on the same extended combustion cylinder in cooperation and in opposite phase.

A 40 horsepower working model of the engine was created and successfully subjected to extensive tests. The automobile manufacturers or others who might have proceeded with further development showed little interest, however.

While in the midst of giving demonstrations, Fessenden was offered and accepted a position by the Submarine Signal Company of Boston, as will be related in the next section. Its management decided that a 500 horsepower version might serve as a marine engine and arranged to produce a larger working model which became operable in 1916. Experiments with it went very well, but they also demonstrated that more research and development

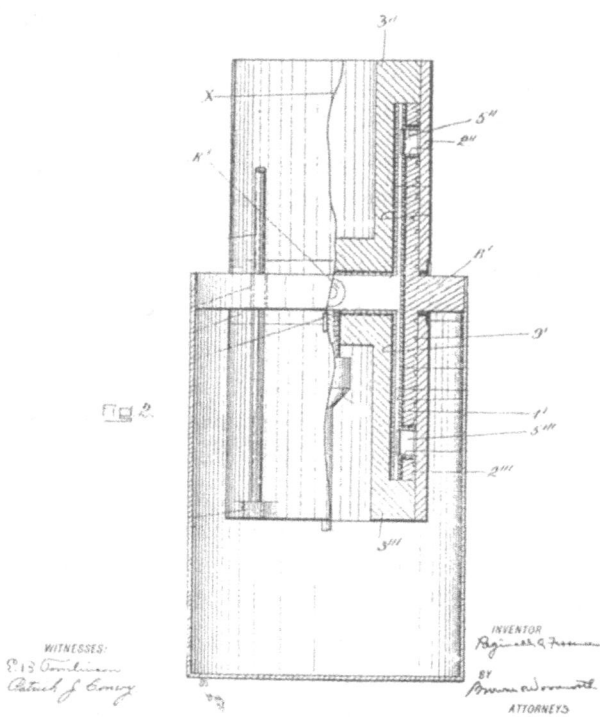

Figure 10) Schematic view of Fessenden's internal combustion engine. The wavy line X divides the elevation view on the left from the cutaway on the right. Shown are two pistons (3'' and 3''') operating in opposite directions on the same combustion cylinder (1'). The closely fitting sleeves which slide on the outside of the combustion cylinder and are attached to the base of the pistons are indicated by 2'' and 2'''. The numbers 5'' and 5''' indicate openings or ports to the inside of the combustion chamber. K' is a duct for distributing fuel to the system. 9' indicates one of the lubricating slots on the exterior of the combustion cylinder. R' is a flange on the interior of an outside casing which supports the combustion cylinder. In the situation shown, the heads of the two pistons have come together near the duct K' to the point of maximum compression. If fuel were contained in the cylinder this would be the point for combustion to occur, driving the two pistons apart. A portion of a drive shaft or connecting rod is shown. It is connected to the exterior sleeve of one of the pistons at one end and to a drive shaft at the other.

would be required before the larger version could be commercialized. The critical problems that remained were related to gas flow, that is, injection of fuel and removal of exhaust gases. Diversions created by the outbreak of World War I, combined with a lack of interest by the U. S. Navy, prevented the company from going further.

In 1921 Fessenden noted that a German firm had picked up work on the larger engine where he had left off. It is very likely that this transfer of interest influenced the design of the original engine for the Volkswagen. The Austrian designer-engineer Ferdinand Porsche (1875-1951), who developed the Volkswagen as well as many other famous automobiles and related devices, took an early interest in light air-cooled engines. It would be difficult to trace any influence Fessenden's work may have had on him, but one can be certain that Fessenden's patents and test-models were noted.

As early as 1911 Fessenden advocated that passenger automobiles be streamlined to cut down air resistance. To facilitate this change in design he also suggested that the engine could be placed in the rear of the vehicle. Nearly twenty years passed before the streamlined Chrysler *Airflow* appeared on the scene. It was a technical success, setting the stage for the future, but was a commercial failure because its appearance seemed too unconventional at the time.

Since the world supply of petroleum seemed to be very limited at the time of World War I, he experimented with the use of naturally available fuels, such as alcohol, in internal combustion engines and concluded that they might prove to be more economical than gasoline in the long run. That day is not yet with us.

The Submarine Signal Company

Fessenden's period of readjustment to his separation from the National Electric Signal Company finally came to an end in the summer of 1912 when he met an old acquaintance from the management of the Submarine Signal Company in Boston. The disaster experienced by the *Titanic* the previous spring had stirred up much public attention and the company was exploring ways in which to improve its ability to detect icebergs, as well as improve signaling more generally. The organization felt that its technology for transmitting signals through the air by a bell was satisfactory, but the microphones used for reception were inadequate. He visited the company offices and was invited to suggest improvements in the latter area.

In this connection it may be noted that in January of 1909, three years before the disaster involving the Titanic, two ships, the *Republic* bound for Europe and the *Florida* for New York, collided off Nantucket in a heavy fog.[14] The former operated by the White Star Line sank slowly. A total of about 1,500 passengers were on the two ships. A dauntless wireless operator in the sinking vessel, Jack Binns of the Marconi Company, stayed at his post for many hours in the bitter cold and succeeded in attracting a third vessel also equipped with wireless to the scene of the accident where it took on a substantial number of the passengers. Binns was properly hailed as a hero internationally. The few lives lost in the incident were an immediate result of the collision. Unfortunately, no steps were taken at that time to make wireless mandatory on shipboard. The *Republic* carried it mainly as a novel luxury item that permitted the passengers to send and receive personal messages when near shore.

In keeping with his vision and talents, Fessenden immediately saw the problem of ship-to-ship communication by sound in a different light: Transmission and reception should use underwater sound waves and adapt the principles employed in wireless transmission to the new medium. Within three months he had developed an oscillator which operated at one kilohertz and was able to produce underwater sound signals that could be detected over a distance of fifty miles by a matching receiver. The road to underwater telegraphy, telephony and sonar had been opened! Needless to say, a long period of testing and development lay ahead.

His high-powered oscillator (figure 11) consisted of a copper tube, about 20 centimeters in diameter and length, placed in a special pattern of strong, constant magnetic fields perpendicular to its axis. The tube was also subject to oscillatory magnetic fields provided by an abutting pair of external coaxial solenoidal coils which induced circumferential currents in the tube. The

---

[14] A detailed account of the event is given in a documentary directed by Dan Loderman and aired in New York over channel 13 of Public Television on February 15, 1999. The title of the documentary is *Rescue at Sea*.

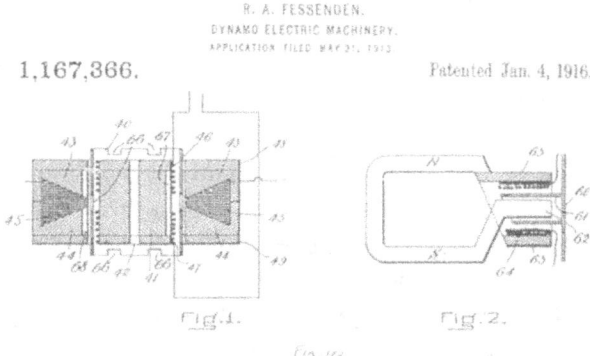

Figure 11). A portion of a page from Fessenden's patent for means of generating powerful undersea acoustical waves. Figure 1 displays the arrangement for setting a copper tube 40 into longitudinal vibration. Shown are two pairs of electromagnets, 43-44, one on either side of the tube. They operate at constant magnetic flux. Their two exciting electric coils are designated 45. The two segments 43 have the same magnetic polarity, for example north, whereas the opposite pair, 44, have the opposite polarity. 41 is a divided iron core or armature located axially within the copper tube 40 as shown. Attached to it are solenoidal coils 46 and 47 through which oscillating currents are passed. The steady magnetic flux from the poles of magnets 43 pass into the armature at the top and out to their respective poles 45 at the bottom. The copper tube is free to move axially. When oscillating currents of a given frequency are passed through the solenoids, they generate circumferential alternating currents of the same frequency in the copper cylinder. These currents interact in turn with the static magnetic fields and exert forces on the copper tube that produce mechanical vibrations in the axial direction. If the solenoids are wound in the same direction and are in series, the ends of the tube will move in opposite directions. An alternative arrangement will generate oscillations in which the ends move in the same direction. When in service, one end of the copper tube is covered with a diaphragm and placed in a liquid medium whose acoustic properties match those of sea water in order to have a good acoustical link with the latter. Figure 2 in the diagram shows an electromagnetic telephone receiver based on the same principles as those used in the sound generator. Here 60 is a diaphragm which is set into motion by a sound wave. Item 61, attached to the diaphragm, is the equivalent of the copper tube in Figure 1 in the diagram; NS is the permanent magnet and 62 is the equivalent of the core or armature. 63 is a circular magnetic pole to which the solenoidal windings 64 are attached. The longitudinal vibration of the tube 61 induces electrical currents in the solenoid. For modern versions of sonic generators based on the use of piezoelectric crystals see, for example, the book *Elements of Acoustical Engineering* by Harry F. Olson, Van Nostrand Company, New York, 1957, currently owned by John Wiley and Sons.

surrounding solenoids, which extended over a distance slightly shorter than the copper tube, were not in contact with the latter, leaving the tube free to move. The forces on the copper tube resulting from the interaction of the steady transverse magnetic fields with the oscillating circumferential currents in it set the tube into longitudinal oscillation with the same frequency as the currents in the solenoids. In this mode, the tube could produce very strong sound waves in the ocean when provided with an appropriate diaphragm and a matching acoustic link to the sea water. He obtained excellent results when the system was fastened to the interior of the external ship plate.

In his patent (U. S. 1,167,366, 1916), he also describes a magnetic microphone based on the same principle that can be used generally for reception. It is not clear whether this was the first practically usable magnetic microphone.

By the winter of 1913-1914, he was prepared to ask for help from Washington to make tests of reflections of underwater sound from icebergs. The tests were started in April of 1914 and carried on into the summer off the Grand Banks. Since icebergs are irregular in shape and the velocity of sound in ice is close to that in sea water, the experimenters observed relatively weak diffuse reflections rather than direct ones. Moreover, it was necessary to use continuous rather that pulsed waves since reflections were observed sporadically. In the best experiments, they picked up the reflections at distances of two and one-half miles.

It was found during the course of these experiments that one could obtain excellent reflections from the ocean floor using a pulsed system. This permitted the development of a device (figure 12), later designated the *Fathometer*, to measure the depth of the water directly under the ship. He obtained a patent on an improved, practical model after the war, encountering the usual legal problems that arise when others attempt to manufacture, market or use a patented device without a license. The U. S. Navy not only used it after World War I without paying appropriate royalties, but its publicity agents credited its invention to the Navy.

To the credit of the U. S. Navy, a later generation of historians and officers did its best to pay homage to Fessenden in its own special way. A World War II destroyer was named after him.

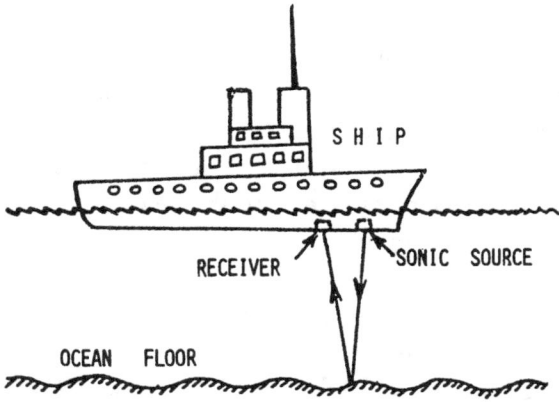

Figure 12. Artist's conception of the operation of the Fathometer, used for measuring the depth of water under the ship. Its operation depends upon measuring the time required for the echo from the sea bottom of a transmitted sound pulse to return to the ship. Fessenden also patented a closely related system for use in geological exploration. See the book mentioned in the caption to figure 10.

World War I.

The First World War broke out in August of 1914, just as the tests with underwater sound were under way. Experiments on the detection of submarines, which had barely started, immediately became the main focus of attention. Since the sympathies of the management of the signaling company were with the Allies, they made immediate contact with the British and Canadian governments with offers to help. They also hoped for a market for their newly developed equipment. Helen Fessenden's biography makes it clear that her husband abhorred the prospect of war, but felt understandably that he must do what was possible to assist his own people through a disastrous period.

Fessenden and his wife sailed to England in the early autumn. He was well received there by the officials to whom he offered exclusive rights to the undersea equipment. They in turn expressed much interest in carrying out tests. The equipment arrived from United States after some delay, at which point the Admiralty stated that it would only be tested for use in communications, not for detecting submarines. Apparently members of the original staff of the Submarine Signal Company had provided a British team with a poorly planned and executed demonstration of the equipment off the U. S. Coast, with the result that it was given a low rating. The company staff had ignored the very precise instructions they had been provided before Fessendan left for England. As an end result, the British failed to show much interest in the new opportunity for locating submarines at that time. In contrast, the German government, which still had access to U. S. inventions and equipment, purchased several units of the undersea devices and adapted them more or less successfully for its own uses.

Fessenden remained in England until the end of 1914, offering advice to various military offices. For example, he proposed the use of sound-ranging to locate hidden guns; he offered Britain the rights to use his light-weight internal combustion engine, developed before joining the submarine company. His most dramatic proposal, however, was that the British build ten thousand airplanes that could carry four 100-pound bombs and use them in massive waves to destroy German factories. Manufacturing could take place in Canada using methods of mass production developed in the United States. He expressed the view that the plan could be put in operation by the middle of 1915 if set into action immediately. The proposal left most individuals who heard it somewhat dazed. H. G. Wells publicized the plan a year or so later, but it was not executed until World War II.

When he returned to Boston and noted what had been occurring there, Fessenden felt that it might be necessary for him to leave the company. Fortunately the management decided to reorganize the staff and operations in ways that he could accept so he returned to his work. A period of relative prosperity for the company ensued as developments and tests continued. The British Admiralty finally showed interest in employing the navigational aids and began to order equipment. Other companies that had a variety of activities began to cross-license the patents for special purposes. The possible use of undersea sound waves as carriers for telephone communication was tested. The system employed was found to be effective for distances as great as five miles in the first experiments. Outside interest in the equipment developed for submarine detection remained relatively dormant however until the United States entered the war in April of 1917.

More Frustrations

In World War II, President Roosevelt created the Office of Scientific Research and Development and the associated National Defense Research Committee, staffed them with some of the best minds in science and engineering available and gave them all the authority and funds needed to support activities that would help win the war. The procedure, perhaps recommended to the President by Churchill, produced miracles.

A much different method of providing scientific and technical guidance was used in World War I. Instead, advisory boards, committees and panels—not uncommonly composed of individuals from industry—were established in individual fields of endeavor and given substantial authority to determine the directions in which endorsement and other support would be provided. The Special Board on Submarine Devices, attached to the Navy Department, was placed under the chairmanship of Willis R. Whitney, Director of the General Electric Research Laboratory. One of the early acts of the board was to establish a research and testing laboratory at Nahant near Boston under the management of the Submarine Signal Company. Both the General Electric Company and the Western Electric Company provided research teams there.

One might have thought that Fessenden, with his extended, successful experience in the field, would have been placed on the board, or at least consulted frequently with respect to research and development in the area to which he had added so much and was prepared to do much more. The opposite was the case. While he remained director in name of the company laboratory, he was completely excluded from any knowledge of the functions of the board or any work that it sponsored. Moreover he was not permitted to guide any of the activity at Nahant. The field of antisubmarine research was closed off to him. It would be difficult at this late date to provide an accurate explanation for the actions of the board; but judging by the standards normally maintained in United States at comparable levels during World War II, they appear to be those of insecure individuals. Fessenden was undoubtedly strong-willed and held strong opinions. However, his ability to make major contributions could not be doubted.

I came to know Willis Whitney reasonably well during a period at the General Electric Research Laboratory in the 1930s. He was then Director

Emeritus, carrying out research of his own choosing with enthusiasm and vitality on matters such as the life history of turtles, the mysteries of the build-up of arterial plaque and the potential use of fractional crystallization for the purification of water. Moreover, he enjoyed discussing his work with any listener who would take the time. It is difficult to believe that he was personally an obstructionist. Another much younger member of the laboratory with whom I worked closely, Clifton Found, had been a member of the General Electric staff at Nahant. He spoke often of the counterproductive barriers that were set up between groups working there and believed that the management of the Western Electric Company had insisted on the isolation of groups.

Fortunately for Fessenden's future reputation, the British, aided by members of the U. S. Navy who were willing to work behind the back of the Special Board on Submarine Devices, were at that time carrying out research and development on submarine detection with considerable success along the lines he started. The war came to a close, however, before their work went much beyond the testing stage. More open research carried out by the U. S. Navy after the war demonstrated clearly that Fessenden's equipment was the best available.

Although Fessenden was disappointed by the treatment he received from the Board on Submarine Devices, he continued to seek out wartime problems on which he could be of service. For example, he invented signaling equipment that would make it possible for a submerged American or British submarine to indicate its nationality if located by an undersea sound signal transmitted from a friendly ship or airplane. He also helped develop flares that could be sent to the surface by a submerged submarine in distress and fired in the air. He worked with the U. S. Secret Service to develop equipment that made it possible to listen to conversations carried on behind thick walls by filtering out noise associated with the natural frequencies of vibration of the room and walls.

It is not surprising that the trials and disappointments to which Fessenden had been subjected since leaving Pittsburgh in 1900 began to take a significant toll on his health, granting that in the meantime he had probably become significantly overweight. In 1916 he had a serious attack of "indigestion" after strenuous exercise. It was then observed that he had abnormally high blood pressure. At the age of fifty he had entered a period of notable and continuing physical decline, including periods of severe angina. Unfortunately the understanding and treatment of coronary disease and arteriosclerosis at that time left much to be desired, compared to present-day standards. Figure 13 shows a portrait-sketch probably made at about this time.

Geophysical Prospecting

In addition to the eventual settlement with RCA, another happy event in Fessenden's business relations occurred in the 1920s. Applying for a patent for the Fathometer in 1917, Fessenden also applied for a patent for the use of sound-generating equipment in geophysical prospecting. Eventually he noted that the Geophysical Research Corporation was employing his procedure in its explorations without taking account of his rights. In this case the company agreed to a settlement without the need for expensive litigation.

Philosophical Issues

During the post-war years, Fessenden spent much of his time reflecting and writing on issues that arose partly from his own experience and partly from speculations of events in ancient history, the latter being supported by a vivid imagination. Those that emerged from his personal experience have timeless interest and readily find a place in the standard literature of our time. His most far-reaching speculations, based on his own interpretation of ancient history, were usually published privately and in some cases distributed with church literature.

He realized that he was living through one of the great historical periods of invention and hoped that its constructive phases would continue indefinitely since invention was doing so much to improve the lot of mankind, initially in the developed countries. He believed that continuation was feasible; the process of invention comes naturally to mankind, granting that the aptitude to invent is not uniformly distributed among individuals. He also realized that the emergence of such periods of creative invention have been sporadic and was inclined to believe that negative forces, such as those that had produced the obstructive actions of wartime boards, must have been responsible for the

Figure 13. A mid-life sketch of Fessenden, from the period of World War I. The artist was his namesake Raymond Aubrey probably a relative on the Trenholme side of the. family. The picture appears in the book dealing with the history of the Fessenden family in America (reference 1). We are very grateful to the editor, Mary Fessenden Washburn, for the privilege of using it. This portrait must have been Helen Fessenden's favorite since it is the only pictorial representation of him in her biography (reference 2).

less creative intervals. His more highly speculative publications involve the suggestion that in times past leading groups possessing much authority had intentionally suppressed invention in some societies.

As might be expected, he felt very strongly that the individual inventor, such as Edison, was the primary source of invention. He doubted if the large industrial laboratories that were being created or expanded in the 1920s could play as significant a role at the frontier of invention; activities would tend to be guided by corporate managerial staff with short-range vision rather by the natural tendencies of the scientists and engineers. We have learned since his time that such laboratories can be highly productive if the individual members of the research staff are given sufficient freedom and support, but his point is well made, particularly at present when so many industrial laboratories are focusing on relatively short-term goals and hoping that research carried on in the universities and other not-for-profit organizations will generate results that prove useful in the long run.

In Fessenden's day, it was not uncommon to believe that, other things being equal, ethnic factors were more important than cultural ones in determining the pace at which invention occurs in a given society. While he was by no means dogmatic, he was inclined to support such a point of view, but with the proviso: "...today we do not know whether a supposed characteristic, e.g. inventive ability, is a real characteristic or due to circumstances." Actually, we now know that cultural factors are the important ones: creativity in places such as Silicon Valley depends on contributions from talented individuals having highly diverse ethnic origins.

He became much interested in the legend of the great flood for which Noah constructed the Ark. He found that the legend of a flood existed in the lore of several ethnic groups that occupied areas surrounding the Black Sea and concluded that it had probably been caused by a massive land slide which generated an enormous tidal wave in what was then a much larger sea. His theory is embellished with many other suggestions that are probably fanciful. Today, it seems most likely that the flood occurred when the Mediterranean Sea broke through the Bosporus into the general area now occupied by the Black Sea, a theory not entirely remote from his.

The "Research Council"

Apparently out of bitterness arising from his treatment during the war, he conjured up an ongoing organization which he termed the "Research Council" which he believed was attempting to gain control of the products of all research carried out in the country. It involved industrial organizations, the private foundations and others not well defined, all of which were combining their efforts toward the common goal. Doubtless the creation of the patent pool linked to the Radio Corporation of America helped fuel his imagination. In any event he believed that all research sponsored by the federal government should become publicly available to any users. He feared that universities would become captives of the great trust and that protective legislation was needed. He also recommended that private foundations be given limited lifetimes, stating that: "These endowments are not, in perhaps most cases, now being used for the purposes which the founders had in mind." Above all, the Courts must protect us from these evils: "No doubt courts are occasionally fallible, as is every human institution, but they are a stronghold of civilization".

In this connection, it should be mentioned that in 1915, before the United States entered World War I, the National Academy of Sciences in Washington established a dependent wing called *The National Research Council*, which was designed to create committees that would provide advice on matters of research and development related to military affairs. Participation would not be restricted to members of the Academy and usually was not. Its status in relation to the federal government was somewhat informal at its start, funding being provided by private sources in keeping with President Wilson's "preparedness" speech of that year. However, it soon gained formal recognition by Presidential Order as a part of the Academy's advisory structure, and prominently remains so up to the present time. A number of the advisory committees created during World War I were initiated through the National Research Council and operated under it. As a result, the term "Research Council" became popularized and used generally to describe such advisory bodies, whether attached to the Academy or not. Fessenden employs the term in this more general sense. The Special Board on Submarine Devices, which caused him so much anguish, was attached directly to the Navy and not to the Academy. Doubtless some of the Academy's committees of the period were equally tendentious.

In his general complaint about the "Research Council," Fessenden makes the following statement:

" Of course, the failure of the Research Council, after the expenditure of so many millions, has to be covered up by propaganda. The following is a list of inventions which the Research Council has prominently published as its work:

Inductor compass for airplanes.
Ultraviolet light signaling
Under-water wireless.
Audion oscillator.
Sonic depth finder.
Wireless compass.
Liberty Motor.
Submarine detector.
Airplane wood-dryer.
Wireless transmission of pictures.
Ultra-audible sound signalling.

but each of which was to my personal knowledge communicated to the Research Council officials by the inventors, e.g.

Inductor compass by Pickard.
Ultraviolet light signaling by Louis Bell.
Audion oscillator by DeForest.
Under-water wireless by Rogers.
Liberty motor by an engineer of the Packard Co. etc. etc.

in each case after practical and successful tests, many of which I witnessed myself, e.g. I operated Louis Bell's ultra-violet signaling apparatus perfectly over a distance of five miles. And in each case the Research Council, after witnessing the tests and privately constructing duplicate models of the inventor's apparatus, arranged that the U.S. Departments should procure the apparatus from members of the Research Cabriri, without paying the inventor anything, and propagandized at government expense, with statements that these inventions were due to the Council."

Aphorisms[2]

Fessenden enjoyed creating aphorisms. A few of his follow:

"All our civilization is based on invention; before invention, men lived on fruits and nuts and pine cones and slept in caves."

"And invention must still go on for it is necessary that we should completely control our circumstances. It is not sufficient that there should (only)* be organization capable of providing food and shelter for all and organization to effect its proper distribution."

"An inventor is one who can see the applicability of means to supply demand five years before it is obvious to those skilled in the art."
"...in going over the history of all the inventions for which history could be obtained it became more and more clear that in addition to training and in addition to extensive knowledge, a natural quality of mind was also necessary."

"The inventor and the research man are confused because they both examine results of physical or chemical operations. But they are exact opposites, mirror images of one another. The research man does something and does not care (exactly)* what it is that happens, he measures whatever it is. The inventor wants something to happen, but does not care how it happens or what it is that happens if it is not what he wants."

"No organization engaged in any specific field of work ever invents any important development in that field, or adopts any important development in that field until forced to do so by outside competition."

(In relation to business)*: "Invention must be its keynote —a steady progression from one new thing to another. As each in turn approaches a saturated market, something new must be *produced*.
"*Personality and salesmanship* do not produce except in a competitive sense.

"*Standardization* does not produce although admirable as an efficiency method.

"*Combination* does not produce though mergers and combinations are still the accepted panacea. In Big Business there appears to be increasing aridity, bureaucracy, and stultifying sacrifice of initiative and above all fear."
(*Note: Items in parenthesis have been introduced by the writer.)

Commentary

The available record gives us some insight into the personality and character of Helen and Reginald Fessenden. As commented earlier, her biographical account of their lives together makes it clear that she was fully as remarkable as he in her own way, even though she tends to mute her role in the text. At one point she amusingly refers to herself as "the cook" of their complex partnership. Actually one can seriously doubt that his accomplishments would have been nearly as extensive as they were if he had not had her as a highly cooperative and far more than competent colleague. It took great courage, foresight and true pioneering spirit to leave the relative security of academic life and plunge into the complex adventures they were to face on a primitive island down the Potomac River, all because he had an inspired vision of the "right" pathway along which the as yet undeveloped field of wireless communications should travel.

Most of the inner workings of the mind of a genius such as Fessenden are beyond normal comprehension. We do know however, that his mind was never at rest and that he rapidly found a new area in which to use his extraordinary gifts if another became blocked. Beyond this, it is also amply clear that he had high ethical standards and was likely to expect the same attribute in those upon whom he depended, often to his discomfort or loss. His published comments[2] with respect to Marconi's "exaggerations" reveal what he expected of others.

He undoubtedly could be very obstinate, an exceedingly important attribute for an individual who would face as many difficult problems as he encountered. One example of his display of temperament is revealing. A prominent engineering society awarded him what was claimed to be a gold medal. On examining it, he found that the gold involved was merely a plated layer on a silver base. He decided that all previous versions of the medal, including one given to Marconi, had been of pure gold and that the society had decided that a plated one would be good enough for him. He returned the medal with a flourish, accepting it a second time only when the president of the society, an old and admired friend, Greenleaf Pickard, assured him that all versions of the medal awarded in the past had also been of the plated variety.

In his essay *"Who Was Fessenden"* [3] George Elliott (figure 14) has the following to say:

"What did Fessenden look like and how did he act? To answer this question we have selected the period of his life from 1903 to 1911 when he was part owner and General Manager of the National Electric Signaling Company (NESCO). Imagine a man distinctly different from his contemporaries, large in body and girth, well over six feet tall, with ginger-colored hair and beard, occasionally wearing a flowing black cape on his shoulders, topped with a seafarer's cap on his head.

"With his razor-sharp mind, his attempts to try to command all situations, his use of his classical scholarship, his lack of patience with slow minds, his restlessness and probing vitality, he seemed easily a character out of a Victorian novel. Chomping on his ever-present cigar, he would argue with one and all on any subject.

"He was sometimes on the verge of poverty and, to continue his experiments, had to negotiate the purchase of railway tickets, lodging, clothes, supplies, etc. at discount prices. He wore the mantle of being an EE professor at two U.S. universities with great pride. Although his time as a faculty member lasted only eight years, he continued to be addressed as "Professor" by associates for the balance of his life.

Figure 14). George Elliott, (Life Member IEEE), retired operations manager engaged in the design, manufacture and installation of gas turbine plants for electric power generation, gas and oil pumping, etc. across Canada. He was one of the founders of the Amateur Radio Fessenden Society and the only editor of the Society's publication *The ARFS Bulletin* until it was terminated in 1996. I am deeply indebted to him for much valuable advice and constructive criticism in the preparation of this manuscript. I am also indebted to John W. Coltman, retired from the former Westinghouse Research Laboratory, who called my attention to George Elliott and the publications of the Fessenden Society.

"Some of his friends and adversaries reaped financial benefit from his early inventions until the last few years of his life when pressure from lawsuits forced the Radio Trust (AT&T, GE, Westinghouse, RCA, et al.) to make some recompense. Through working as a consultant during his post-NESCO years, he earned substantial sums of money, particularly for his detection and earth-strata measuring patents. One may speculate on the notion that if a novelist created a book character like RAF, it would have been considered improbable.

"While overly aggressive and perhaps domineering, Fessenden had some endearing qualities. Dr. Alexanderson, developer of GE's RF alternator, said he was a charming and kind person, and was forthright in business matters. Some insight into RAF's tendency to exhibit compassion is revealed by his solicitude for his cat, "Mikums," especially at the time of its sickness and death.

"He was successful in earning the loyalty of his associates and employees."

Above all, Fessenden was an idealist and a visionary much concerned about human destiny. He believed that our species could and should achieve much higher levels of civilization and enlightenment. He maintained that two of the essential requirements for attaining such levels are continuing, acceptable refinement of the laws which govern society and never-ending encouragement of the process of invention in order to optimize use of the opportunities nature offers us. He devoted much of his life, brilliantly, to pursuit of the second, using the best tools available to him. While most individuals are inclined to regard his contributions to the development of radio as by far the most important, I would place his work on sonar at a comparable level.

NOTE: In addition to individuals mentioned in the text, I would like to express gratitude to the following persons who were of substantial help in preparing this biographical essay: First, Mrs. Florence Arwade who manages my office and kept everything on track in addition to providing additional extraordinary service, and Ms. Patricia. E. Mackey, the University Librarian, who went to great effort to locate many out of the way sources of books and papers. Then I am indebted for special technical advice to John Blewett of Chapel Hill, North Carolina, John S. Coleman of Arlington, Virginia, H. Richard Crane of Ann Arbor, Michigan, Norman G. Einspruch of Coral Gables, Florida, Herbert Friedman of Arlington, Virginia, Kenneth G. McKay of Chapel Hill, North Carolina, William Teare of Aurora, Colorado, Joachim Seitz of Palo Alto, California, and Robert N. Varney of Palo Alto, California. I am also indebted to Mrs. Lucie Marshall of Gualala, California and Mrs. Ruth Brown of San Francisco, California for many valuable suggestions with respect to the text. Finally, I would like to express gratitude to the editorial staff of the American Philosophical Society for its excellent advice and cooperation.

REGINALD AUBREY FESSENDEN

HIS MIND
ILLUMINED THE PAST
AND THE FUTURE
AND WROUGHT GREATLY
FOR THE PRESENT

BY HIS GENIUS
DISTANT LANDS CONVERSE
AND MEN SAIL
UNAFRAID UPON
THE DEEP

HELEN MAY FESSENDEN

HIS WIFE

A NOBLE COURAGEOUS AND
BRILLIANT WOMAN WHO GAVE
INSPIRATION TO ALL WHO KNEW HER

(Inscriptions on the tomb of Reginald and Helen Fessenden in Bermuda and shown on the back cover)

APPENDIX I

PUBLICATIONS OF REGINALD FESSENDEN
(BASED ON LISTING IN HELEN FESSENDEN'S BIOGRAPHY)

SECTION I
WIRELESS TELEGRAPHY, TELEPHONE AND
TELEPHOTOGRAPHY

Lodge Wave Telegraphy, *Electrical World and Engineer* July 29, 1899.
Lodge Wave Telegraphy, *Electrical World and Engineer*, Aug. 12, 1899.
Wireless Telegraphy, *Electrical World and Engineer*, Sept. 16, 1899.
The Possibilities of Wireless Telegraphy, *American Institute of Electrical Engineering* Nov. 22, 1899.
Wireless Telegraphy over Frozen Ground, *Electrical World and Engineer*, Jan.26, 1901.
Wireless Telegraphy, *Electrical World and Engineer*, June 27, 1901.
The Relative Reliability of Wireless and Wire Telegraph systems, *Electrical World and Engineer*, Nov. 14, 1903.
Collins Articles, *Electrical World and Engineer*, Aug. 23, 1902.
Collins Articles, *Electrical World and Engineer*, Sept. 19, 1903.
Theories in Wireless Telegraphy, *Electrical World and Engineer*, Jan. 13, 1904.
The Government Use of Wireless Telegraphy, *Electrical World and Engineer*, Aug. 20, 1904.
Wireless Telegraphy, *The Electrician*, Sept. 16, 1904.
Wireless Telegraphy, *The Electrician*, Feb. 3, 1905.
Water-Stream Antenna, *Electrotechnischen Zeitschrift*, Nov. 6, 1905.
Water-Stream Antenna, *Electrotechnischen Zeitschrift*, Feb. 6, 1906.
Water-Stream Antenna, *Electrotechnischen Zeitschrift*, July 19, 1906.
Wireless Telegraphy, *Electrical Review*, May 11, 1906.
Wireless Telegraphy, *Electrical Review*, May 18, 1906.
Interference in Wireless Telegraphy and the International Telegraph Conference, *Electrical Review*, July 6, 13, 20, 27, 1906.
Austin Thermo-Electric Wave Detector, *Electrical World*, Nov. 10, 1906.
The Wireless Telegraph Situation, *Scientific American*, Jan. 19, 1907.

APPENDIX I: FESSENDEN PUBLICATIONS

Recent Progress in Wireless Telephone, *Scientific American*, Jan. 19, 1907.
The Continuous Production of High Frequency Oscillations, *The Electrician*, Feb. 15, 22, 1907.
Wireless Telephony, *Electrical Review*, Feb. 15, 22 and March 1, 1907.
The Principles of Electric Wave Telegraphy, *The Electrician*, July 5, 1907.
The Principles of Electric Wave Telegraphy, *The Electrician*, Sept. 13, 1907
Wireless Telegraphy During Daylight, *The Electrician*, July 26, 1907.
Atmospheric Absorption of Wireless Signals, *Electrical Review*, Sept 6, 1907.
Wireless Telegraphy, *Scientific American*, Sept. 28, 1907.
Long Distance Wireless Telephony, *The Electrician*, Oct. 4, 1907.
A Regular Wireless Telegraph Service between America and Europe, *Scientific American Supplement*, Nov. 16, 1907; *The Electrician*, Nov. 22, 1907; *Electrical Review*, Nov. 22, 1907.
Trans-Atlantic Wireless Telegraphy, *Engineering*, Jan. 18, 1907.
Trans-Atlantic Wireless Telegraphy, *Engineering*, Jan. 25, 1907
Trans-Atlantic Wireless Telegraphy, *The Electrician*, Jan. 3, 1908.
Wireless Telegraphy, *Electrical Review*, Jan. 17, 1908.
Portable Type of High-Frequency Alternator, *The Electrician*, July 3, 1908.
The Predetermination of the Radiation Resistance of Antennae, *The Electrician*, Aug. 7, 1908.
Wireless Telephony, *The Electrician*, Nov. 27, 1908.
Tantalum Wave Detectors and Lamps, *the Electrician*, Feb. 5, 1909.
Correspondence with Reference to Obtaining Permit from Colonial Office to Provide the West Indies and Canada with Cheaper Telegraphic Communication, *British Blue Book*, July, 1910.
Statement Submitted, *British Blue Book*, July, 1910.
Memorandum, Proposals, Etc., *British Blue Book*, July, 1910.
How Ether Waves Really Move, *Popular Radio*, November, 1923.

SECTION II
SUBMARINE TELEGRAPHY AND TELEPHONY

Long-Distance Submarine Signalling by Dynamo-Electric Machinery, *American Academy of Arts* and the *Lawrence Scientific Association* in joint session: Boston, Feb. 25, 1914.

The Fessenden Pelorus (Wireless Compass), A caution as to its use, *Electrician*, Dec. 19, 1919.

SECTION III
WIRE TELEGRAPHY, TELEPHONY AND CABLES

Sine Form of Curves of Alternating E.M.F., *Electrical World*, New York, Sept. 15, 1894.

Sine Form of Curves of Alternating E.M.F., *Electrical World*, New York, Sept. 29, 1894.

The Cause of Change of Microphone Resistance, *American Electrician*, Feb., 1897.

Microphonic Telephonic Action, *American Electrician*, May, 1897.

Electromagnetic Mechanism, with Reference to Telegraphic Work, *Journal of the Franklin Institute*, June, 1900.

SECTION IV
PHOTO-BOOK

Use of Photography in Data Collections, *Electrical World*, New York, Aug. 22, 1896

SECTION V
ELECTRICAL ENGINEERING

Non-Arcing Metals, *Electrical Engineer*, New York, Apr. 6, 1892.

Vacuum Tube Lightning Arresters, *Electrical Engineer*, New York, Aug. 17, 1892.

Fireproof Insulation, *Electrical World*, New York, Sept. 3, 1892.

Conductors and Insulators, *Electrical World*, New York, Mar. 18, 1893.

Conductors and Insulators-II, *Electrical World*, New York, Mar. 18, 1893.
Conductors and Insulators-III, *Electrical World*, New York, Mar. 26, 1893
Conductors and Insulators-IV, *Electrical World*, New York, May 6, 1893
Conductors and Insulators-V, *Electrical World*, New York, May 13, 1893.
Conductors and Insulators-VI, *Electrical World*, New York, May 20, 1893.
Boilers for Small Central Stations, *Electrical World*, New York, Mar. 3, 1894.
Steam Boilers for Central Stations, *Electrical World*, New York, Mar. 10, 1894.
Anthony on the Incandescent Lamp, *Transactions of the American Institute of Electrical Engineers*, Vol. XI, March, 1894.
Definition of a Polyphase System, *Electrical World*, New York, Mar. 30, 1895.
On the Relation between Maximum Induction and Remanance, *Electrical World*, New York, Aug. 3, 1895.
The Loss of Energy in Changing from a Single Alternating Current to Polyphase Currents, *Electrical World*, New York, Dec. 7, 1895.
Probable Development in Electricity and Electrical Engineering, *Electrical World*, New York, Mar. 7, 1896.
Economic Use of Electric Power for Driving Tools, *Engineers' Society of Western Pennsylvania*, Sept. 1896.
Some New Electrical Apparatus, *Electrical World*, New York, Dec. 5, 1896.
The Evolution of the Rail Bond, *Electrical World*, New York, Feb. 5, 1898.
The Evolution of the Rail Bond, *Electrical World*, New York Mar. 19, 1898.
The Evolution of the Rail Bond, *Electrical World*, New York, Mar. 23, 1898.
Insulation and Conduction, *American Institute of Electrical Engineering*, Mar. 23, 1898.
The Relation between Mean Spherical and Mean Horizontal Candle Power of Incandescent Lamps, *Electrical World*, New York, Feb. 25, 1899.
Frequency Meters, *Electrical World*, New York Nov. 11, 1899.
The Method of Insulation by Freezing, *Electrical World*, New York, Sept 8,

1900.

Magnetic Observations and Traction Disturbances, *The Electrician*, London, Jan. 11, 1901.

Electrolytic Rectifiers, *Electrical World and Engineer*, June 1, 1901.

Recent Progress in Practical and Experimental Electricity, *The Philosophical Society*, Oct. 12, 1901.

Discussion of D. McFarlan Moore's Paper, *Transactions American Institute of Electrical Engineering*, Apr. 26, 1907.

SECTION VI
GENERATION AND STORAGE OF POWER

A Sun Storage Battery, *American Electrician*, May, 1898.

Official Report of the Ontario Power Commission, Mar. 28, 1906.

The Commercial Solution of the Problem of Utilizing, for the Production of Power, the energy of Solar Radiation, the Wind and other Intermittent Natural Sources, *The Times*, London, Sept. 8, 1910.

"Banking" Electricity for Universal Use, *Scientific American*, April 30, 1921.

Boston May Revolutionize Heating Problem, *Boston Evening Transcript*, Nov. 29, 1922.

Cheaper Electric Heat is Demonstrated Possibility, *Boston Evening Transcript*, Jan. 27, 1926.

SECTION VII
GENERAL PHYSICS AND CHEMISTRY

An Electrically Driven Gyrostat, *Electrical Engineer*, May 19, 1889.

Electricity in Chemical Manipulations, *Chemical News*, London, Jan. 3, 1890.

The Volumetric Analysis of Copper, *Chemical News*, London, Apr. 18, 1890.

The Volumetric Analysis of Copper, *Chemical News*, London, May 23, 1890.

The Setting up of Clark Standard Cells, *Electrical World*, New York, June

APPENDIX I: FESSENDEN PUBLICATIONS

7, 1890.

Action of Nitric Acid on Asphalt and Cellulose, *Chemical News*, London, Mar. 18, 1892.

Electrical Discharge through a Geissler Tube, *Science*, New York, Apr. 21, 1893.

Effect of a Gaseous Envelope on the Resistance of a Metal, *The Electrician*, London, June 30, 1893.

A New Method of Preventing Heat Radiation, *Electrical World*, New York, Jan. 13, 1894.

Standards of Illumination, *Transactions of the American Institute of Electrical Engineering*, Feb. 23, 1894.

Standards of Illumination, *Transactions of the American Institute of Electrical Engineering*, May 21, 1895.

Standards of Illumination, *Transactions of the American Institute of Electrical Engineering*, May 20, 1896.

Standards of Illumination, *Transactions of the American Institute of Electrical Engineering*, June 28, 1899.

Variations in Resistance:

On a Proposed Modification of the Generally Accepted Temperature Co-efficient of Resistance for Copper Wires, *Electrical World*, New York, Feb. 16, 1895.

On the Electrolysis of Gases, *Astrophysical Journal*, Chicago, Dec. 1895.

A New Method of Measuring Temperature, *Nature*, London, Jan. 16, 1896.

Outline of an Electrical Theory of Comets's Tails, *Astrophysical Journal*, Dec. 1896.

The Movement of Encke's Comet, *Nature*, London, Sept. 29, 1898.

On the Use of the Methven Standard with Blackened Chimney, *Electrical World*, New York, Feb. 28, 1899.

Absolute Determination of the Ohm, *Nature*, London, Apr. 271, 1899.

Nature of the Lightning Discharge, *Electrical World* and *Electrical Engineer*, Apr. 29, 1899.

A Multiple Lightning Flash, *Electrical World and Engineer*, Nov. 4, 1899.

The True Explanation of Dark Lightning Flashes, *Electrical World and Engineer*, Jan. 6, 1900.

Physics at the American Association, *Science*, New York, July 20, 1900.
Light without Heat, *Electrical World and Engineer*, Jan. 5, 1901.
India Rubber, *The Electrician*, London, Nov. 6, 1903.
On Thermo-Galvanometers, *The Electrician*, London, June 24, 1904.
On Thermo-Galvanometers (Corrections), *The Electrician*, London, July 15, 1904.
The High-Pressure Electric Condenser, *The Electrician*, London, Nov. 3, 1905.
On the Magnetic Properties of Electrolytic Iron, *Transactions of the American Institute of Electrical Engineering*, May 30, 1906.
Wireless Telegraphy and the Ether, *Eastern Association of Physics Teachers*, Nov. 23, 1912.
A Sage Method of Using Mercury Bichloride for the Antisepsis of Wounds of Large Surface, *Science*, New York, June 18, 1915.

SECTION VIII
MATHEMATICS

The Centimetre Gramme Second and the Centimetre Dyne Second Systems of Units and a New Gravitational Experiment, *Science*, New York, Dec. 22, 1893.
A Formula for the Area of the Hysteresis Curve, *Electrical World*, New York, June 9, 1894.
Magnetic Formulae, *Electrical World*, New York, June 23, 1894.
On the True Dimensions of the Electrostatic and Electromagnetic Units, and on the Right Use of the Terms Intensity, Strength, Force and H, *Electrical World*, New York, May 4, 1895.
The Quantity upon which a Knowledge of the Nature of Electricity and Magnetism Depends, *Electrical World*, New York, May 18, 1895.
Dimension Formulae and the Theory of Units, *Electrical World*, New York, June 29, 1895.
On the Use of Magnetic Formulae in Electrical Design, *electrical World*, New York, Aug. 24, 1895.
Qualitative Mathematics, *Electrical World*, New York, Feb. 6, 1897.
How to get rid of "4" Eruption without changing any of the Legal Units,

Electrical World and Engineer, Dec. 9, 1899. *Electrician*, London, Dec. 29, 1899.

A Proposed System of Units, *The Electrician*, Dec. 29, 1899.

Motion of Committee on Units and Standards, *Proceedings of American Institute of Electrical Engineering*, Mar. 28, 1900.

On a System of Units, *The Electrician*, May 20, 1904

SECTION IX
ECONOMICS

On Professional Degrees, *Electrical World and Engineer*, New York, Nov. 11, 1899.

Colonial Telegraphic Communication, *Times*, London, Oct. 26, 1910.

SECTION X
AGRICULTURAL ENGINEERING

Fessenden Patent, No. 1121722, Dec. 22, 1914.

Fessenden Patent, No. 1268949, June 11, 1918.

SECTION XI
COHESION AND MOLECULAR PHYSICS

Note on the Volume Force of Solids, *Electrical World*, Aug. 8, 1891.

Atomic Volume and Tensile Strength, *Electrical World*, Aug. 22, 1891.

Theory of Solution, *Electrical Review*, London, Nov. 27, 1891

Use of Glucinum in Electrical Instruments, *Electrical World*, New York, July 16, 1892.

The Laws and Nature of Cohesion, *Science*, New York, July 22, 1892, March 3, 1893, *Chemical News*, Oct. 21, 1892, Oct. 27, 1893.

Some Recent Work on Molecular Physics, *Journal of the Franklin Institute*, Sept. 1896.

SECTION XII
NATURE OF ELECTRICITY, MAGNETISM AND GRAVITATION

On the Prospective Development of Ether Theories, *Electrical World*, New York, Jan. 2, 1897.

On the Prospective Development of Ether Theories, *Electrical World*, New York, Jan. 30, 1897.

A Determination of the Nature of the Electric and Magnetic Quantities and of the Density and Elasticity of the Ether, *Physical Review*, Cornell, Jan. 1900.

An Explanation of Inertia, *Electrical World and Engineer*, April 7, 1900.

Inertia and Gravitation, *Science*, New York, Aug. 31, 1900.

As to the Nature of Inertia and Gravitation, *Transactions of the Toronto Astronomical Society*, 1901.

An Explanation of Gravitation, *Electrical World and Engineer*, New York, Sept. 29, 1900.

Theories of Gravitation, *Electrical World land Engineer*, New York, Oct. 13, 1900.

A Determination of the Nature and Velocity of Gravitation, *Science*, New York, Nov. 16, 1900.

Cohesion, Electricity, Magnetism and Gravitation, Unpublished, Written June 1909.

Transformation of Gravitational Waves into Ether Vortices, *Science*, New York, Oct. 17, 1913.

Gyroscopic Quanta, *Science*, New York, April 10, 1914.

Quantum Radiation a Gyroscopic Phenomenon, Unpublished, Written July 26, 1914.

SECTION XIII
HISTORICAL

The Deluged civilization of the Caucasus Isthmus, Privately printed and distributed, also, through Massachusetts Bible Society, 1923.

Finding a Key to the Sacred Writings of the Egyptians, *Christian Science Monitor*, March 18, 1924.

APPENDIX I: FESSENDEN PUBLICATIONS

How it was discovered that all so-called Myth-Lands were the Caucasus Isthmus, *Christian Science Monitor*, March 8, 1926.

Chapter XI—of the Deluged Civilization, Privately printed and distributed; also through Massachusetts Bible society, 1927.

The Founding of Empire Day, Privately printed and distributed 1930.

An apparently definite Identification of Masons with the Egyptian M-S-N, *Mersyside* Association for Masonic Research, 1932.

The Deluged Civilization of the Caucasus Isthmus (unpublished and reprinted papers), Posthumously Published, Privately printed and distributed, 1933.

Appendix II:
Some Major Fessenden Patents

(The numbers are designations given by the U. S, Patent Office)

Silicon Alloys

452,494 (1891)

Design and Construction of X Ray Equipment

648,660 (1900)

Heterodyne Principle

706,738 (1902)

706,739 (1902)

706,740 (1902)

1,050,441 (1913)

1,050,728 (1914)

Rectification

706,736 (1902)

F. Alternator

706,737 ((1902)

Arc Oscillator (Mention)

706,742 (1902)

Radio Telephone

706,747 (1902)

Multiplexing

715,203 (1902)

727-326 (1903)

981,406 (1911)

SOME FESSENDEN PATENTS

Point Contact Rectifier

727,327 (1903)

Electrolytic Detector

727,331 (1903)

Vertical Antenna

793,651 (1905)

Anti Static Device

918,306 ((1909)

918,307 (1909)

Directive Antenna Array

1,020,032 (1912)

Storage of Wheeled Vehicles

1,114,975 (1914)

Internal Combustion Engine

1,132,465 (1915)

Sound Production and Signaling

1,207,387 (1916)

1,207,388 (1916)

1,311,157 (1916)

1,108,895 (1914)

1,277,562 (1918)

1,384,855 (1920)

Submarine Signaling and Detection

1,348,556 (1920)

1,348,828 (1920)

1,348,855 (1920)

1,429,497 (1922)

Subsurface Directive Signaling

1,348,856 (1920)

1,355,598 (1920)

Ship Location

1,319,145 ((1919)

Fathometer

1,217,585 (1917)

Geophysical Prospecting With Sound

1,240,328 (1917)

Water Storage and Power Generation

1,214,531 (1910)

1,247,520 (1917)

Gun Location by Sound

1,341,795 (1920)

Microphotographic Books

1,616,848 (1927)

1,732,302 (1929)

SUBJECT INDEX

alternating current generator, 27, et seq.
    General Electric Company, 27, 29
Amateur Radio Fessenden Society (ARFS), 7
amplitude modulated radio, 1
aphorisms, 541 et seq.
automobile engine, 39
    German interest, 41
    lightweight, 39, et seq.
backroom support of sonar by U.S. Navy, 48
Biblical flood, 51, et seq.
Bishop's College, Quebec, 9
break in health, 1916, 48
Canadian support
    search, 19, 32
Carnegie-Mellon interests, 20
Chemical fire retardant, 1
Clark Collection
    Smithsonian Institution, 7
Cobb Island, 17
    voice transmission, 17, 18
Coherer, 16
Collision of *Republic* and *Florida*, 1909, 42
Columbian Exposition 1893, 12
continuous wave transmission, 15
    modulation, 15

DeVeaux Military College, 8
direct current motors, 11

earphones, 16
Edison Machine Works, 9
electrical conduits
    laying, 9
electrically driven gyro, 1
electrolytic rectifier, 16
electromagnetic theory
    indispensable, 10
Empire Day, 4

fathometer, 44
    Navy use, 44
Ferrosilicon alloys, 1, 12, 13
fireproof electrical insulation

antimony chloride, 10, 15

geophysical prospecting, 1, 49

heterodyne principle, 3
heterodyne system, 26
    oscillating arc, 27
    Poulsen arc, 27, 35
    U.S. Navy, 27, 34, et seq.

International Yacht Race of 1899, 16
    Marconi, 16
    New York Herald, 16
ionosphere, 12

library of Queens University, Ontario, 7
Lick Observatory, 14
light air-cooled engine, 1, 39
litigation
    Fessenden patents, 33, et seq.
Llewellyn Park Laboratory of Edison, 10
    closure, 11

machine gun
    Maxim, 12
Marconi Company, 19, 32
Maxwell's equipment, 13
Maxwell-Hertz Waves, 1, 15
    challenge, 1
    Fessenden, 1
    Marconi, 1
measuring equipment
    Weston, 12
microfilm, 1

National Electric Signaling Company, 19, 20
    collapse, 32
    sale, 33
National Research Council, 1915, 52, et seq.
National Academy of Sciences
    nature, 9
Navy destroyer *Reginald Fessenden*, 44
New York Public Library, 7, 8

Ontario Power Commission, 38
    Fessenden s proposals, 38, et seq.

Patents
    importance, 8
    over 200, 1

patents, De Forest, 31
philosophical issues, 49, et seq.
photography, 15
point-contact diodes, 16
popularity of radio, 36
Principles of Radio Communications, 1921, 35
    Wastershed document, 35
psychological trauma
    Reginald Fessenden, 11
    Reginald Kennelly, 12
Purdue University, 13

Queen and Company of Philadelphia, 20

radar, 37
Radio Corporation of America (RCA), 33
    Settlement, 1928, 33
Radio News, 4
Radio's First Voice, 7
rectification, 16
"Reduction to practice," 3
"Research Council," 52, et seq.

Scientific American, 9
sonar, 1, 42, et seq.
    Fessenden s devices, 42, et seq.
    first magnetic microphone (?), 44
Special Board on Submarine Devices, 47, et seq.
    disaster for Fessenden, 47
    Nahant laboratory, 47
    Willis Whitney, 47, 48
Stanley Company, Pittsfield, 13
State Archives, North Carolina, 7
steam turbine
    Parsons, 13
streamlined automobile, 41
Submarine Signal Company, 41, et seq.
    Titanic disaster, , 41
Superheterodyne system, 36
    AM band 37

E.H. Armstrong, 36
intermediate frequency, 36, 37
litigation, 37
local mixing oscillator, 36, 37
RCA monopoly, 37

television, 37
theoretical chemistry, 10
Tiffany, 11
transatlantic transmission
    accidental voice transmission, 25
    collapse of antenna, 21, 25
    "Echo" signal, 25
    optimum frequencies, 25
    two way, 21, et seq.
Trinity College School, 9
turbo-electric drive, 1, 30
two-way transatlantic wireless, 1

U.S. Weather Bureau, 17
    disaster, 18
    Willis L. Moore, 17
underground telephone, 10
underwater sound, 34
United States Company, Westinghouse, 11
University of Pittsburgh, 14

vacuum tube circuitry, 26
    local oscillator, 26
    oscillator, 35
vacuum tube development
    AT&T, 31
    Langmuir, 31
vacuum tube era, 3
voice Transmission, 26 et seq.
    holiday broadcast, 30

Westinghouse, 11
Whitney Institute, Bermuda, 9
"Who was Fessenden," 56, et seq.
wireless regulation, 30, et seq.
World War I, 45, et seq.
    Aerial bombing, 46
    mixed reception, 46
    visit to England, 45, et seq.

X-ray equipment, 1
X-ray equipment, 14

NAME INDEX

Alexanderson, E. F. W., 29, 58
Armstrong, E. H., 34, 35, 36

Binns, Jack, 42
Bose, J. C., 15
Brashear, John, 14, 27

Curry, W. A., 35

De Forest, Lee, 25, 31
Duddell, William, 2

Edison, Thomas A., 9, et seq., 51
Elliot, George, 4, 7, 11, 20, 56
Ewing, J. A., 13

Fessenden, Cortez Ridley Trenholme, 4
Fessenden, Elisha Joseph, 4
Fessenden, Helen, 33, 45, 55
Fessenden, John, 3
Fessenden, Kenneth Harcourt, 4
Fessenden, Peter, 7
Fessenden, Reginald A., 1, et seq.
Fessenden, Reginald Kennelly, 11
Fessenden, Victor Lionel, 4
Fessenden, William Pitt, 3
Fish, Frederick p., 31
Found, Clifton, 48

Gernsback, Hugo, 4
Given, T. H., 20,

Keeler, James A., 14
Kennelly, Arthur E., 11
Kintner, Samuel M., 32, 37

Langley, Samuel, 147
Langmuir, Irving, 31

Marconi, Guglielmo, 1, 11, 17, 21, 32, 56
Maxim, H. 5., 12
Moore, Willis L., 17, 19
Morecroft, J. H., 35

Parsons, Charles, 13
Pickard, Greenleaf, 56

Pinto, A., 35
Porsche, Ferdinand, 41
Poulsen, 34
Poulsen, Vladimir, 27

Raby, Ormand, 7
Roebling, John, 21
Roentgen, W., 14
Roosevelt, Franklin, D., 47
Roosevelt, Theodore, 19

Stanley, Mr., 13

Thomsom, J. J., 13
Trenholme, Clementina, 4
Trenholme, Edward, 3
Trott, Helen May, 9
Twain, Mark, 25

Vail, Theodore N., 31

Walker, Hay, Jr., 20, 32, 33
Wells, H. G., 46
Westinghouse, George, 12, 14
Weston, J. A., 12
Whitney, Willis R., 47
Wilson, Woodrow W., 52
Wolcott, Darwin 5., 19, 20
Wright, Orville, 18
Wright, Wilbur, 18

Young, Owen D., 33

www.ingramcontent.com/pod-product-compliance
Lightning Source LLC
Chambersburg PA
CBHW080801020526
44114CB00035B/7